Selected Titles in This Series

604 **Hajnal Andréka, Steven Givant, and István Németi,** Decision problems for equational theories of relation algebras, 1997

603 **Bruce N. Allison, Saeid Azam, Stephen Berman, Yun Gao, and Arturo Pianzola,** Extended affine Lie algebras and their root systems, 1997

602 **Igor Fulman,** Crossed products of von Neumann algebras by equivalence relations and their subalgebras, 1997

601 **Jack E. Graver and Mark E. Watkins,** Locally finite, planar, edge-transitive graphs, 1997

600 **Ambar Sengupta,** Gauge theory on compact surfaces, 1997

599 **Tai-Ping Liu and Yanni Zeng,** Large time behavior of solutions for general quasilinear hyperbolic-parabolic systems of conservation laws, 1997

598 **Valentina Barucci, David E. Dobbs, and Marco Fontana,** Maximality properties in numerical semigroups and applications to one-dimensional analytically irreducible local domains, 1997

597 **Ragnar-Olaf Buchweitz and John J. Millson,** CR-geometry and deformations of isolated singularities, 1997

596 **Paul S. Bourdon and Joel H. Shapiro,** Cyclic phenomena for composition operators, 1997

595 **Eldar Straume,** Compact connected Lie transformation groups on spheres with low cohomogeneity, II, 1997

594 **Solomon Friedberg and Hervé Jacquet,** The fundamental lemma for the Shalika subgroup of $GL(4)$, 1996

593 **Ajit Iqbal Singh,** Completely positive hypergroup actions, 1996

592 **P. Kirk and E. Klassen,** Analytic deformations of the spectrum of a family of Dirac operators on an odd-dimensional manifold with boundary, 1996

591 **Edward Cline, Brian Parshall, and Leonard Scott,** Stratifying endomorphism algebras, 1996

590 **Chris Jantzen,** Degenerate principal series for symplectic and odd-orthogonal groups, 1996

589 **James Damon,** Higher multiplicities and almost free divisors and complete intersections, 1996

588 **Dihua Jiang,** Degree 16 Standard L-function of $GSp(2) \times GSp(2)$, 1996

587 **Stéphane Jaffard and Yves Meyer,** Wavelet methods for pointwise regularity and local oscillations of functions, 1996

586 **Siegfried Echterhoff,** Crossed products with continuous trace, 1996

585 **Gilles Pisier,** The operator Hilbert space OH, complex interpolation and tensor norms, 1996

584 **Wayne W. Barrett, Charles R. Johnson, and Raphael Loewy,** The real positive definite completion problem: Cycle completability, 1996

583 **Jin Nakagawa,** Orders of a quartic field, 1996

582 **Darryl McCollough and Andy Miller,** Symmetric automorphisms of free products, 1996

581 **Martin U. Schmidt,** Integrable systems and Riemann surfaces of infinite genus, 1996

580 **Martin W. Liebeck and Gary M. Seitz,** Reductive subgroups of exceptional algebraic groups, 1996

579 **Samuel Kaplan,** Lebesgue theory in the bidual of $C(X)$, 1996

578 **Ale Jan Homburg,** Global aspects of homoclinic bifurcations of vector fields, 1996

577 **Freddy Dumortier and Robert Roussarie,** Canard cycles and center manifolds, 1996

576 **Grahame Bennett,** Factorizing the classical inequalities, 1996

575 **Dieter Heppel, Idun Reiten, and Sverre O. Smalø,** Tilting in Abelian categories and quasitilted algebras, 1996

574 **Michael Field,** Symmetry breaking for compact Lie groups, 1996

(*Continued in the back of this publication*)

Decision Problems for Equational Theories of Relation Algebras

MEMOIRS
of the
American Mathematical Society

Number 604

Decision Problems for Equational Theories of Relation Algebras

Hajnal Andréka
Steven Givant
István Németi

March 1997 • Volume 126 • Number 604 (end of volume) • ISSN 0065-9266

American Mathematical Society
Providence, Rhode Island

1991 *Mathematics Subject Classification.*
Primary 03G15; Secondary 03B25, 03B05, 08B15.

Library of Congress Cataloging-in-Publication Data

Andréka, H.

Decision problems for equational theories of relation algebras / Hajnal Andréka, Steven Givant, István Németi.

p. cm. — (Memoirs of the American Mathematical Society, ISSN 0065-9266 ; no. 604)

"March 1997, volume 126, number 604 (end of volume)."

Includes bibliographical references.

ISBN 0-8218-0595-9 (alk. paper)

1. Relation algebras. 2. Decidability (Mathematical logic) I. Givant, Steven R. II. Németi, I. III. Title. IV. Series.

QA3.A57 no. 604
[QA10]
510 s—dc21
[511.3′24] 96-37450
CIP

Memoirs of the American Mathematical Society

This journal is devoted entirely to research in pure and applied mathematics.

Subscription information. The 1997 subscription begins with number 595 and consists of six mailings, each containing one or more numbers. Subscription prices for 1997 are \$414 list, \$331 institutional member. A late charge of 10% of the subscription price will be imposed on orders received from nonmembers after January 1 of the subscription year. Subscribers outside the United States and India must pay a postage surcharge of \$30; subscribers in India must pay a postage surcharge of \$43. Expedited delivery to destinations in North America \$35; elsewhere \$110. Each number may be ordered separately; *please specify number* when ordering an individual number. For prices and titles of recently released numbers, see the New Publications sections of the *Notices of the American Mathematical Society.*

Back number information. For back issues see the *AMS Catalog of Publications.*

Subscriptions and orders should be addressed to the American Mathematical Society, P. O. Box 5904, Boston, MA 02206-5904. *All orders must be accompanied by payment.* Other correspondence should be addressed to Box 6248, Providence, RI 02940-6248.

Memoirs of the American Mathematical Society is published bimonthly (each volume consisting usually of more than one number) by the American Mathematical Society at 201 Charles Street, Providence, RI 02904-2294. Periodicals postage paid at Providence, RI. Postmaster: Send address changes to Memoirs, American Mathematical Society, P. O. Box 6248, Providence, RI 02940-6248.

This publication is indexed in *Science Citation Index*®, *SciSearch*®, *Research Alert*®, *CompuMath Citation Index*®, *Current Contents*®*/Physical, Chemical & Earth Sciences.*
Printed in the United States of America.

∞ The paper used in this book is acid-free and falls within the guidelines established to ensure permanence and durability.

10 9 8 7 6 5 4 3 2 1 01 00 99 98 97

CONTENTS

ABSTRACT

We prove that any variety of relation algebras which contains an algebra with infinitely many elements below the identity, or which contains the full group relation algebra on some infinite group (or on arbitrarily large finite groups), must have an undecidable equational theory. Then we construct an embedding of the lattice of all subsets of the natural numbers into the lattice of varieties of relation algebras such that the variety correlated with a set X of natural numbers has a decidable equational theory iff X is a decidable (i.e., recursive) set. Finally, we construct an example of an infinite, finitely generated, simple, representable relation algebra that has a decidable equational theory.

TO BJARNI JÓNSSON

INTRODUCTION

The foundation of an algebraic theory of binary relations was laid by C. S. Peirce, building on earlier work of Boole and De Morgan. The basic universe of discourse of this theory is a collection of binary relations over some set, and the basic operations on these relations are those of forming unions, complements, relative products (i.e., relational compositions), and converses (i.e., inverses). There is also a distinguished relation, the identity relation. Other operations and distinguished relations studied by Peirce are definable in terms of the ones just mentioned. Such an algebra of relations is called a *set relation algebra.*

A modern development of this theory as a theory of abstract *relation algebras*, axiomatized by a finite set of equations, was undertaken by Tarski and his students and colleagues, beginning around 1940. In 1942, Tarski proved that all of classical mathematics could be developed within the framework of the equational theory of relation algebras. Indeed, he created a general method for interpreting into the equational theory of relation algebras first-order theories that are strong enough to form a basis for the development of mathematics — in particular, set theories and number theories. As a consequence, he established that the equational theories of relation algebras and of set relation algebras are undecidable (see Tarski [1941], p. 88 and Tarski-Givant [1987], Theorem 8.5(xii)(β) and the historical remark in Footnote 3* on p. 168). They were the first known examples of undecidable equational theories. As was pointed out in *op. cit.*, Tarski's proof actually shows more. Namely, any class of relation algebras that contains the full set relation algebra on some infinite set (i.e., the set relation algebra whose universe consists of all binary relations on some infinite set) must have an undecidable equational theory.

Tarski continued to be interested in decision problems for classes of relation algebras, and at various times he posed one or more such problems to his students. We gather together here some of these problems.

TARSKI'S DECISION PROBLEMS FOR RELATION ALGEBRAS. *Which of the following classes have undecidable equational theories?*

(1) *Group relation algebras, i.e., algebras of complexes (subsets) of groups under the usual Boolean and group complex operations.*
(2) *Abelian group relation algebras.*
(3) *Boolean group relation algebras.*
(4) *Abelian relation algebras, i.e., relation algebras in which the relative product operation is commutative.*
(5) *Abelian set relation algebras.*
(6) *Symmetric relation algebras, i.e., relation algebras in which conversion is the identity operation.*

Received by the editor February 7, 1995.

(7) *Symmetric set relation algebras.*
(8) *Finite relation algebras.*
(9) *Finite set relation algebras.*

As a consequence of the first theorem below, we conclude that each of these classes has an undecidable equational theory.

The pervasiveness of undecidable theories within the lattice of equational theories of relation algebras leads to the problem of determining when such theories are decidable. Of course, it is trivial to observe that the equational theory of a finite relation algebra (or, equivalently, of any finite collection of finite relation algebras) is decidable. It is natural to ask whether these are the only examples. Our second theorem establishes a strong negative answer to this question. Namely, there is a complete embedding f of the lattice of subsets of the set ω of natural numbers into the lattice of equational theories of relation algebras such that, for each $X \subseteq \omega$, $f(X)$ is a decidable theory iff X is a decidable (i.e., recursive) set. Moreover, none of the theories $f(X)$ is the theory of a finite algebra.

To obtain the theories in the range of f, we begin with a class of finite relation algebras exhibiting a great deal of structural uniformity. (These are the well-known Lyndon algebras constructed from projective lines.) We group the algebras together in recursive and non-recursive ways to arrive at (classes of algebras with) decidable and undecidable equational theories.

Because these theories are constructed using uniformly definable finite algebras, they exhibit certain strong properties, such as local finiteness, i.e., their finitely generated models are all finite. The question arises whether this is just a peculiarity of our construction or whether it is a necessary attribute of such decidable theories. In other words, are there examples of decidable equational theories of relation algebras that are not locally finite?

In the final chapter of this work we construct an example of an infinite set relation algebra with one generator whose equational theory, though far from trivial, is still decidable. This result is summarized in our third theorem below.

We shall say that an equational theory Ψ of relation algebras is *recursively inseparable* if there is no recursive set of equations that contains the equational theory Φ of relation algebras and is disjoint from $\sim\Psi$, the set of equations not in Ψ (i.e., the set of equations falsifiable in models of Ψ); equivalently, Ψ is recursively inseparable if there is no recursive set Θ of equations such that $\Phi \subseteq \Theta \subseteq \Psi$. In particular, if Ψ is recursively inseparable, then every equational theory of relation algebras included in Ψ is undecidable.

THEOREM. *Let* K *be a class of relation algebras satisfying one of the following two conditions*:

(i) *For each* $n \in \omega$, *there is a simple algebra in* K *with at least* n *elements below the identity*;
(ii) *For each* $n \in \omega$, *there is an algebra in* K *containing a set* G *with at least* n *elements such that* G *is a group under the relative product and converse operations*, *and the elements of* G *are pairwise disjoint.*

Then the equational theory of K *is recursively inseparable.*

Rather surprisingly, the requirement in condition (ii) that G be a group under the relative product and converse operations can be weakened: it is sufficient (and indeed equivalent) to require that G be *closed* under the relative product and converse operations and that G contain an equivalence element. In this alternate formulation it is essential to retain the requirement that the elements of G be pairwise disjoint.

In practice, condition (i) and condition (ii) apply in rather different situations, as we shall see below. It is therefore surprising that any class of relation algebras which satisfies condition (i) must also satisfy condition (ii). In other words, we can really delete condition (i) from the statement of the previous theorem, keeping only (ii). However, condition (i) is often encountered in concrete applications and it is easy to check.

Any class K of relation algebras that contains, e.g., the full group relation algebra on an infinite group or on arbitrarily large finite groups will satisfy condition (ii); indeed, the atoms of a full group relation algebra form a group under the relative product and converse operations.

COROLLARY. *If* K *is a class of relation algebras satisfying one of the two conditions in the preceding theorem, then* K *has an undecidable equational theory.*

Tarski's original theorem (in the stronger form mentioned above) follows at once from the corollary, using either condition (i) or (ii). Indeed, the full set relation algebra on an infinite set has infinitely many elements below the identity, and contains the full group relation algebra on an infinite group as a subalgebra. In a similar fashion, we conclude that the class of all finite relation algebras, the class of all finite set relation algebras, and the class of all set relation algebras on finite sets have undecidable equational theories. The equational undecidability of this last class was originally established in Schönfeld [1979].

Condition (i) fails for classes of relation algebras in which the simple algebras have few elements below the identity, for example classes of relation algebras in which the simple algebras are integral, i.e., the relative product of non-zero elements is non-zero. Each of (1)–(7) is such a class. However, using condition (ii) we can conclude, e.g., that each of the following classes has an undecidable equational theory: Abelian relation algebras, Abelian set relation algebras, symmetric relation algebras, symmetric set relation algebras, integral relation algebras, integral set relation algebras, group relation algebras, Abelian group relation algebras, and Abelian p-group relation algebras, for any prime p (in particular, Boolean group relation algebras). So does each of the corresponding subclasses of finite algebras, and the classes of Abelian, symmetric, and integral set relation algebras on finite sets. And so do the full group relation algebras on the additive groups of the integers, the rationals, the reals, and the complexes. (The undecidablity of the equational theory of symmetric relation algebras is due to Jipsen [1992], who first pointed out that it can be derived rather easily from the main theorem of Maddux [1981]. See the end of Chapter 2 for further historical remarks, and in particular for remarks regarding the history of the methods used to prove the above theorem.)

Recall that the lattice of equational theories of relation algebras is dually isomorphic to the lattice of universal classes of simple relation algebras (see the preliminaries). Thus, any result about the former can be formulated in terms of the latter. Let L be the class of subalgebras of Lyndon algebras (on sets of cardinality at least 3). We shall show in Chapter 3 that L is a universal class of simple relation algebras. Thus, the set of universal subclasses of L forms a sublattice of the lattice of all universal classes of simple relation algebras. For each $X \subseteq \omega$, let L_X be the class of algebras $\mathfrak{A}$ in L such that, for all $n \in \omega \sim X$, $\mathfrak{A}$ is not isomorphic to the Lyndon algebra on a set of cardinality $n+3$. Then L_X is a universal subclass of L.

THEOREM. *The mapping* $X \longmapsto \mathsf{L}_X$ *is a complete embedding of the lattice of subsets of* ω *into the lattice of universal subclasses of* L. *The equational theory of* L_X *is decidable iff* X *is decidable. Further, for no* X *is the equational theory of* L_X *the theory of a finite algebra. Finally, every universal subclass of* L *that is not in the range of this embedding has a decidable equational theory.*

Thus, for example, the equational theory of the class L_ω, i.e., of L itself, is decidable and is not the theory of a finite algebra.

COROLLARY. *There is no smallest variety of relation algebras with an undecidable equational theory.*

This corollary solves for relation algebras an open problem that was first formulated by Németi [1985] for finite dimensional cylindric algebras and later for relation algebras.

It turns out that the class L, and in fact $\mathbf{SP}(\mathsf{L})$ and each of its subclasses, is locally finite. The next theorem shows that there are interesting non-locally finite relation algebras with decidable equational theories.

THEOREM. *There is an infinite set relation algebra that is not locally finite* (*in fact, it is generated by one element*) *and yet has a decidable equational theory.*

The results announced above have been applied in Kurucz-Németi-Sain-Simon [1993] to solve problems in the domain known as Logic, Language and Information. In particular, they have been used to answer questions raised by van Benthem [1994] and Pratt [1994] concerning Arrow Logic and fragments of Linear Logic (cf. Marx-Masuch-Polos [a]). They have also been used to investigate certain types of varieties of Boolean algebras with operators; see, for example, Németi-Sain-Simon [1995].

ACKNOWLEDGEMENTS. The results of this paper were announced in Andréka-Givant-Németi [1994a], [1994b]. The research of Andréka and Németi was supported by the Hungarian National Foundation for Scientific Research, Grant Nos. T16488 and T7255. Givant's research was supported by grants from the International Research and Exchanges Board (IREX) and from Mills College.

CHAPTER I

PRELIMINARIES

We suppose the ordinal numbers to be defined so that each ordinal ξ is the set of all of its predecessors, i.e., $\xi = \{\eta : \eta < \xi\}$. In particular, $n = \{0, \ldots, n-1\}$ for every natural number n. The set of all natural numbers will be denoted by ω. The greatest integer less than or equal to a given real number x is denoted, as usual, by $\lfloor x \rfloor$ and the least integer greater than or equal to x by $\lceil x \rceil$.

For each set X, we write $|X|$ for the cardinality of X, and $Sb\,(X)$ for the collection of all subsets of X. Given sets X and Y, we denote their set-theoretic difference by $X{\sim}Y$, i.e., $X{\sim}Y = \{x \in X : x \notin Y\}$; when X is a fixed universe of discourse, we write "$\sim Y$" instead of "$X{\sim}Y$". We denote the Cartesian product of X and Y by $X \times Y$, i.e., $X \times Y = \{(x, y) : x \in X \text{ and } y \in Y\}$. The set of functions from X to Y is denoted by ${}^XY$. If f is a function whose domain X and range Y include U and V respectively, then $f[U]$ is the image of U under f, i.e., $\{f(u) : u \in U\}$, and $f^{-1}[V]$ is the inverse image of V under f, i.e., $\{u \in X : f(u) \in V\}$. If $a = \langle a_\xi : \xi < \kappa\rangle$ and $b = \langle b_\eta : \eta < \lambda\rangle$ are two (ordinally indexed) sequences, then $a\hat{\ }b$ denotes their concatenation, i.e., the sequence c with domain $\kappa+\lambda$ such that $c_\xi = a_\xi$ when $\xi < \kappa$, and $c_\xi = b_\eta$ when $\xi = \kappa + \eta$ and $\eta < \lambda$.

We shall identify binary relations on a set U with subsets of $U \times U$, and we shall usually use the letters P, Q, R, S, T, possibly subscripted, to denote binary relations. The *relative product* or *relational composition* of two binary relations R and S, and the *converse*, or *inverse* of such a relation are defined in the usual way:

$$R|S = \{(x, y) : \text{there is a } z \text{ such that } (x, z) \in R \text{ and } (z, y) \in S\}\,,$$
$$R^{-1} = \{(x, y) : (y, x) \in R\}\,.$$

The *identity* and *diversity* relations on U will be denoted by I_U and D_U respectively, or simply by I and D, when no confusion can arise. Notice that, for functions F and G from U to U, the functional composition $F \circ G$ is not the same as the relational composition $F|G$; in fact, $F|G = G \circ F$.

Capital German letters will denote algebras and the corresponding Roman letters their universes. For example, A is the universe of $\mathfrak{A}$. Occasionally, following standard practice, we shall refer to A as if it were the algebra. For instance, when dealing with a group $\mathfrak{G}$, we may talk about a subset H of G being a subgroup. An algebra is said to be *trivial* if its universe has only one element and *non-trivial* otherwise. The *reduct* of an algebra $\mathfrak{A} = \langle A\,,\, O_\xi\rangle_{\xi \in \Xi}$ to a subset Γ of its operations (more precisely, to a subset Γ of its index set) is the algebra $\mathfrak{A}' = \langle A\,,\, O_\xi\rangle_{\xi\in\Gamma}$.

Given a class K of algebras, the class of all algebras isomorphic to homomorphic images, subalgebras, or direct products of systems of members of K is denoted by $\mathbf{H}(\mathsf{K})$, $\mathbf{S}(\mathsf{K})$, or $\mathbf{P}(\mathsf{K})$ respectively. An algebra is κ*-generated* if it is generated by some set of cardinality at most κ, and *finitely generated* if it is κ-generated for some finite cardinal κ. An algebra is *locally finite* if every finitely generated subalgebra is finite. A class of algebras is said to be locally finite if every member of the class is locally finite.

We assume that a first-order language is correlated with each class of algebras of a given similarity type. A formula in such a language is *open* if it has no quantifiers. An open formula that is the disjunction of one equation with a (possibly empty) collection of negations of equations is called a *conditional equation* or *quasi-equation* or *strict universal Horn formula*. The universal closure of an open formula, a conditional equation, and an equation are called, respectively, a *universal sentence*, a *strict universal Horn sentence* or *quasi-identity*, and an *identity*. The class of all models (in a given similarity type) of a set of formulas (or, equivalently, a set of sentences), a set of open formulas (or, equivalently, a set of universal sentences), a set of conditional equations, and a set of equations, is called, respectively, an *elementary class*, a *universal class*, a *strict universal Horn class* or *quasi-variety*, and an *equational class* or *variety*. The set of all open formulas, all conditional equations, and all equations true of a class K of algebras is called the *open* or *universal theory*, the *conditional equational* or *strict universal Horn theory*, and the *equational theory* of K. The equational theory of K is denoted by $\mathcal{E}q(\mathsf{K})$. (A similar terminology and notation apply to various types of theories of a single algebra.) By a celebrated theorem of Birkhoff [1935] (in a form due to Tarski [1946]), the variety generated by K, i.e., the class of models of $\mathcal{E}q(\mathsf{K})$, is just $\mathbf{HSP}(\mathsf{K})$.

Suppose $\tau(v_0,\dots,v_{n-1})$ (with variables among $v_0,\dots,v_{n-1}$) and $\sigma_0,\dots\sigma_{n-1}$ are terms in the language of some algebra $\mathfrak{A}$. Then $\tau(\sigma_0,\dots\sigma_{n-1})$ is the term obtained from τ by simultaneously substituting $\sigma_0,\dots\sigma_{n-1}$ for the variables $v_0,\dots,v_{n-1}$. $\tau^{\mathfrak{A}}$ denotes the n-ary operation on A induced by τ. Strictly speaking, this notation should also contain a reference to the rank n of the operation, for example $\tau^{\mathfrak{A}}_n$. However, it will always be clear from our discussion which n we have in mind. We shall omit similar remarks in other, analogous situations. If a is an n-termed sequence of elements from $\mathfrak{A}$, then we write $\tau^{\mathfrak{A}}[a]$ or simply $\tau[a]$ for the value of $\tau^{\mathfrak{A}}$ at a. Suppose now that the variables of τ are among $v_0,\dots,v_{n-1},w_0,\dots,w_{m-1}$. When concentrating on the variables $v_0,\dots,v_{n-1}$, we may regard the variables $w_0,\dots,w_{m-1}$ as *parameter variables*. If b is an m-termed sequence from A, then $\tau^{\mathfrak{A}}[v_0,\dots,v_{n-1},b]$ denotes the n-ary operation on A induced by τ with the *parameters* b; when the sequence of parameters is clear, we shall often omit any reference to it and write simply $\tau^{\mathfrak{A}}$.

As usual, we write $\mathfrak{A}\models\varphi[a]$ to express the fact that the n-termed sequence a from $\mathfrak{A}$ satisfies the formula $\varphi(v_0,\dots,v_{n-1})$ in $\mathfrak{A}$. $\varphi^{\mathfrak{A}}$ is the set of all such sequences; thus, it is an n-ary relation on A. Suppose that φ has (parameter) variables $w_0,\dots,w_{m-1}$ in addition to the variables $v_0,\dots,v_{n-1}$. If b is an m-termed sequence from A, then $\varphi^{\mathfrak{A}}[v_0,\dots,v_{n-1},b]$ denotes the n-ary relation on A defined by φ with the parameters b, i.e., $\{a\in {}^nA : \mathfrak{A}\models\varphi[a\hat{\ }b]\}$. Again, when the sequence of parameters is clear, we shall omit any reference to it and write simply $\varphi^{\mathfrak{A}}$.

For the next few paragraphs, we fix an algebra

$$\mathfrak{A}=\langle A\,,+\,,-\,,;\,,\smile\,,1'\rangle$$

of type $\langle 2,1,2,1,0\rangle$, i.e., $+$ and ; are binary operations on A, $-$ and $\smile$ are unary operations, and 1' is a distinguished element. The first-order language appropriate for this type of algebra shall be denoted by $\mathcal{L}_r$. We take 1, 0, and 0' to be, respectively, the constants defined by

$$1 = 1' + -1' \quad , \quad 0 = -(1' + -1') \quad , \quad 0' = -1' \quad ,$$

and $\cdot$ and $\leq$ to be, respectively, the binary operation and binary relation on A defined by

$$x \cdot y = -(-x + -y) \quad \text{and} \quad x \leq y \text{ iff } x + y = y \, .$$

We shall employ the following exponential notation for the operation ; :

$$x^0 = 1' \quad , \quad x^1 = x \quad , \quad \text{and} \quad x^{n+1} = x^n \, ; x \quad \text{for } n \geq 1 \, .$$

Whenever parentheses indicating the order of performance of the operations are lacking, it will be understood that (1) each of the unary operations $-$ and $\smile$ has priority over the binary operations, (2) the multiplications $\cdot$ and ; have priority over addition $+$, (3) ; has priority over $\cdot$, (4) in case (1)–(3) are not applicable, the operations should be performed from left to right.

The algebra $\mathfrak{A}$ (referred to above) is a *relation algebra* if it satisfies the following (equationally expressible) conditions: $\langle A \, , + \, , - \rangle$ is a Boolean algebra, $\langle A \, , ; \, , {}^\smile \, , 1' \rangle$ is an involuted monoid, ; and $\smile$ are distributive operations over $+$, and we have $x^\smile \, ; -(x \, ; y) \leq -y$ for all $x, y \in A$. Thus, the class RA of all relation algebras is a variety, i.e., an equationally axiomatizable class of algebras. The postulate $x^\smile \, ; -(x \, ; y) \leq -y$ may seem rather complicated and unintuitive. In the presence of the other axioms, it is equivalent to the following *cycle law*:

$$x \cdot (y \, ; z) \neq 0 \quad \text{iff} \quad y \cdot (x \, ; z^\smile) \neq 0 \quad \text{iff} \quad z \cdot (y^\smile \, ; x) \neq 0 \, .$$

When x, y, z are atoms, the cycle law can equivalently be formulated as follows:

$$x \leq y \, ; z \quad \text{iff} \quad y \leq x \, ; z^\smile \quad \text{iff} \quad z \leq y^\smile \, ; x \, .$$

We shall refer to $\langle A \, , ; \rangle$ as the *relative product* reduct, and to $\langle A \, , ; \, , {}^\smile \, , 1' \rangle$ as the *Peircean* reduct, of $\mathfrak{A}$. The constants 1' and 0' are called the *identity* and *diversity* elements of $\mathfrak{A}$.

The sum or supremum of a set X of elements of a relation algebra $\mathfrak{A}$ is denoted by $\sum^{\mathfrak{A}} X$, or simply by $\sum X$, provided that this sum exists. Notice that $\sum \varnothing = 0$. The operations $\smile$ and ; are *completely distributive* over Boolean addition in the following sense: for all subsets X and Y of A such that $\sum X$ and $\sum Y$ exist, the sums

$$\textstyle\sum\{x^\smile : x \in X\} \quad \text{and} \quad \sum\{x \, ; y : x \in X \, , \, y \in Y\}$$

exist, and

$$\begin{aligned} (\textstyle\sum X)^\smile &= \textstyle\sum\{x^\smile : x \in X\} \, , \\ (\textstyle\sum X) \, ; (\textstyle\sum Y) &= \textstyle\sum\{x \, ; y : x \in X \, , \, y \in Y\} \end{aligned}$$

(see Chin-Tarski [1951], Theorems 1.1 and 2.3). A (unary) function f with domain A is *additive* if $f(x + y) = f(x) + f(y)$ for all x, y in A, and *completely additive* if $f(\sum X) = \sum f[X]$ for all $X \subseteq A$ such that $\sum X$ exists. In particular, if $\sum X$ exists, then $\sum f[X]$ is assumed to exist. The notions of additivity and complete additivity can also be extended to functions of higher rank (see Jónsson-Tarski [1951], Definition 1.1). Thus, we can also say that the operations $\smile$ and ; are completely additive.

The algebra $\mathfrak{A}$ is a *set relation algebra* if A is a collection of binary relations over some *base set* U, and $+$, $-$, ; , and $\smile$ are the ordinary set-theoretic operations on binary relations of union, complementation with respect to the unit $\bigcup A$, relative product, and conversion; furthermore, 1' is the identity relation on U. It is easy to check that, in general, $\bigcup A$ is some equivalence relation on U. When $\bigcup A = U \times U$, we say that $\mathfrak{A}$ is a *standard* set relation algebra *on* U (this terminology is not standard), and when A is the collection of all binary relations on U, i.e., $A = Sb(U \times U)$, we say that $\mathfrak{A}$ is the *full* set relation algebra on U. The latter is denoted by $\mathfrak{R}(U)$. Its relative product reduct, $\mathfrak{S}(U) = \langle Sb(U \times U), | \rangle$, is a semigroup that will play an important role in our work.

A relation algebra is *representable* if it is isomorphic to a set relation algebra. RRA is the class of all representable relation algebras. Tarski [1955] showed that RRA is a variety, while Monk [1964] proved that it is not finitely axiomatizable.

$\mathfrak{A}$ is a *group relation algebra* on a group $\mathfrak{G} = \langle G, \circ, {}^{-1}, e \rangle$ if A is a collection of subsets, or *complexes*, of G, while $+$ and $-$ are the set-theoretic operations on complexes of union and complementation with respect to G, ; and $\smile$ are the group-theoretic operations of multiplication and inversion of complexes, i.e.,

$$H ; K = \{h \circ k : h \in H \text{ and } k \in K\} \quad \text{and} \quad H^{\smile} = \{h^{-1} : h \in H\},$$

and 1' $= \{e\}$. When the underlying group is, e.g., an Abelian group, a p-group, or a Boolean group, then we shall speak of an *Abelian* group relation algebra, a p-group relation algebra, or a *Boolean* group relation algebra, respectively. If $A = Sb(G)$, then $\mathfrak{A}$ is called the *full* group relation algebra on $\mathfrak{G}$ or the *complex algebra* of $\mathfrak{G}$; we shall denote this algebra of all complexes by $\mathfrak{Cm}(\mathfrak{G})$. Given an infinite group $\mathfrak{G}$, it is not difficult to show that the collection of finite and cofinite subsets of G forms the universe of a subalgebra of $\mathfrak{Cm}(\mathfrak{G})$ (see Andréka-Givant-Németi [1994], Lemma 20). We denote this subalgebra by $\mathfrak{Cf}(\mathfrak{G})$, and call it the *finite-cofinite* group relation algebra on $\mathfrak{G}$. We shall be particularly interested in group relation algebras on the following additive groups: the integers, $\mathfrak{Z}$; the integers modulo n, $\mathfrak{Z}_n$; the rationals, $\mathfrak{Q}$; the reals, $\mathfrak{R}$; and the complexes, $\mathfrak{C}$.

Using the Cayley representation of groups as groups of permutations, it is easy to see that a group relation algebra on $\mathfrak{G}$ is isomorphic to a standard set relation algebra on G. Indeed, for each $g \in G$, put

$$g^* = \{(h, h \circ g) : h \in G\}.$$

Then the mapping

$$H \longmapsto \bigcup\{h^* : h \in H\}$$

is an embedding of $\mathfrak{Cm}(\mathfrak{G})$ into $\mathfrak{R}(G)$. (It is essential here that we use right, not left, multiplication by g, since the operation of composition in $\mathfrak{R}(G)$ is relational, not functional, composition.) The mapping $g \longmapsto g^*$ takes $\mathfrak{G}$ isomorphically to the subalgebra

$$\mathfrak{G}^* = \langle \{g^* : g \in G\} \, , \, | \, , \, ^{-1} \, , \, I_G \rangle$$

of the Peircean reduct of $\mathfrak{R}(G)$. Notice that this isomorphism takes distinct elements to pairwise disjoint elements. For suppose that $f, g \in G$ and that $f^* \cap g^* \neq \varnothing$. Let $h, k \in G$ be such that

$$(h, h \circ f) \in f^* \quad , \quad (k, k \circ g) \in g^* \quad , \quad (h, h \circ f) = (k, k \circ g) \, .$$

Then $h = k$ and $h \circ f = k \circ g$, so $h \circ f = h \circ g$ and therefore $f = g$.

Tarski [1955] showed that the class GRA of algebras isomorphic to group relation algebras is a universal class, i.e., it is axiomatizable by a set of universal sentences.

Several special kinds of relation algebras and relation-algebraic elements will play an important role in our discussion. An element x of a relation algebra is a *subidentity* element if it is below 1', a *subdiversity* element if it is below 0', *functional* if $x^{\smile} ; x \leq 1'$, *permutational* if $x^{\smile} ; x = 1'$ and $x ; x^{\smile} = 1'$, *reflexive* if $1' \leq x$, *symmetric* if $x^{\smile} = x$, *transitive* if $x ; x \leq x$, and an *equivalence element* if it is symmetric and transitive. As is shown in Chin-Tarski [1951], Theorem 3.2, x is an equivalence element iff $x ; x = x = x^{\smile}$ iff $x ; x^{\smile} = x$. Two elements x and y *commute* if $x ; y = y ; x$, and are *disjoint* if $x \cdot y = 0$. A relation algebra is *atomic* if it is non-trivial (i.e., if it has more than one element) and if every non-zero element is above some atom, *symmetric* if every element is symmetric, and *Abelian* if any two elements commute. A symmetric relation algebra is always Abelian.

Recall that an algebra is *simple* if it has just two congruence relations. (This condition is equivalent to the following requirements: the algebra is non-trivial and all non-constant homomorphisms over the algebra are in fact embeddings). An algebra is *subdirectly irreducible* if it has a (unique) smallest congruence relation that is different from the identity relation. (This condition is equivalent to the following requirements: the algebra is non-trivial; furthermore, whenever an embedding of the algebra into a product of algebras has the property that its composition with the projections maps *onto* the factor algebras, then the composition with at least one of the projections is an isomorphism). As is well known, a relation algebra is simple iff it is subdirectly irreducible. Thus, every relation algebra is *semi-simple*, i.e., a subdirect product of simple relation algebras (see Jónsson-Tarski [1952], Theorems 4.14 and 4.15). In particular, an equation or a conditional equation holds in all relation algebras iff it holds in all simple relation algebras. Tarski [1955] proved that every representable relation algebra is also a subdirect product of simple representable relation algebras. Thus, an equation or a conditional equation holds in all representable relation algebras iff it holds in all simple representable relation algebras. Jónsson-Tarski [1952], Theorem 4.28, proved that a representable relation algebra $\mathfrak{A}$ is simple iff it is embeddable in $\mathfrak{R}(U)$ for some non-empty set U, i.e., iff it is isomorphic to a standard set relation algebra on a non-empty set. If such an embedding exists for which U is finite, respectively infinite, then $\mathfrak{A}$ is said to be *finitely*, respectively *infinitely*, representable. Of course, a relation algebra may be both finitely and infinitely representable. Also, a finite relation algebra need not be finitely representable. A well-known example is the eight-element set relation algebra over the set of rational numbers that is generated by the relation $<$. McKinsey and Tarski proved that a relation algebra is simple iff the sentence

$$0 \neq 1 \wedge \forall x(x \neq 0 \rightarrow 1 ; x ; 1 = 1)$$

or, equivalently, the sentence

$$0 \neq 1 \wedge \forall x \forall y(x \neq 0 \wedge y \neq 0 \rightarrow x ; 1 ; y \neq 0)$$

is true of it (see Jónsson-Tarski [1952], Theorem 4.10). Thus, for any variety V of relation algebras, the class $\mathbf{Si}(\mathsf{V})$ of simple algebras in V is a universal class. In particular, the class of all simple representable relation algebras is universal.

A relation algebra is *integral* if it is non-trivial, and the relative product of two non-zero elements is always non-zero. As Jónsson-Tarski [1952], Theorem 4.17, points out, a relation algebra is integral iff 1' is an atom. Of course, an integral relation algebra must be simple (see the second displayed formula above). Although the converse fails in general (an example is the full set relation algebra on a set with at least two elements), a simple Abelian relation algebra is always integral. Indeed, suppose that $\mathfrak{A}$ is simple, but not integral — say x and y are disjoint, non-zero, subidentity elements. Then $(x ; 1) ; y \neq 0$, since $\mathfrak{A}$ is simple, and $y ; (x ; 1) = 0$, since x and y are disjoint subidentity elements. Therefore $x ; 1$ and y do not commute, so $\mathfrak{A}$ cannot be Abelian. Since a symmetric relation algebra is Abelian, it follows that a simple symmetric relation algebra is integral.

With every open formula φ in the language of relation algebras, we can correlate in an effective way an equation ε_φ with the same free variables such that the equivalence $\varphi \leftrightarrow \varepsilon_\varphi$ is universally valid in all simple relation algebras. Indeed, the equation $\sigma = \tau$ is equivalent to the equation $\sigma \cdot \tau + -\sigma \cdot -\tau = 1$. Thus, without loss of generality we may assume that each equation and inequality has "1" as its right-hand term. The inequality $\sigma \neq 1$ is equivalent to the equation $1 ; -\sigma ; 1 = 1$ in all simple relation algebras, and the conjunction $(\sigma = 1) \wedge (\tau = 1)$ is always equivalent to the equation $\sigma \cdot \tau = 1$. From these observations, using an induction based on the inductive definition of open formulas, it is easy to define the desired correlation. (This correlation essentially goes back to Schröder; see, e.g., Jónsson [1982] or Tarski-Givant [1987], p. 26.) We state a simple conclusion of this *open formula reducibility* that is known from the literature (see Jónsson [1982], pp. 284, 285, and 290, or else Andréka-Givant-Németi [1994], Theorem 1).

THEOREM 1.1. (i) *For every variety* V *of relation algebras,* $\mathbf{Si}(\mathsf{V})$ *is a universal class, and for any universal class* K *of simple relation algebras,* $\mathbf{SP}(\mathsf{K})$ *is a variety. Moreover,* $\mathbf{Si}(\mathbf{SP}(\mathsf{K})) = \mathsf{K}$ *and* $\mathbf{SP}(\mathbf{Si}(\mathsf{V})) = \mathsf{V}$. *Therefore, the correspondence* $\mathsf{V} \longmapsto \mathbf{Si}(\mathsf{V})$ *maps the complete lattice of varieties of relation algebras isomorphically onto the complete lattice of universal classes of simple relation algebras. Its inverse is the correspondence* $\mathsf{K} \longmapsto \mathbf{SP}(\mathsf{K})$.

(ii) *In the lattice of universal classes of simple relation algebras, the join and meet of two classes are respectively the union and intersection of the classes.*

Notice that when V is the variety of trivial algebras, then $\mathbf{Si}(\mathsf{V})$ is the empty universal class.

COROLLARY 1.2. *Let* $\mathfrak{A}_0, \mathfrak{A}_1, \ldots, \mathfrak{A}_{n-1}$ *be finite, simple relation algebras and* K *the variety generated by these algebras, i.e.,* $\mathsf{K} = \mathbf{HSP}(\{\mathfrak{A}_0, \ldots, \mathfrak{A}_{n-1}\})$*. Then*

$$\mathbf{Si}(\mathsf{K}) = \mathbf{S}(\{\mathfrak{A}_0, \ldots, \mathfrak{A}_{n-1}\}),$$

i.e., the simple algebras in K *are just the algebras that are embeddable into one of* $\mathfrak{A}_0, \ldots, \mathfrak{A}_{n-1}$*. In particular, up to isomorphisms* K *has only finitely many simple algebras.*

PROOF. Since the algebra $\mathfrak{A}_i$ is finite for each $i < n$, it can be completely described by a first-order sentence φ_i. Because n is finite, the class of isomorphic copies of $\mathfrak{A}_0, \ldots, \mathfrak{A}_{n-1}$ is an elementary class — in fact, the class of models of $\bigvee_{i<n} \varphi_i$. Therefore, by the Łoś-Tarski Theorem (see, e.g., Tarski [1954], Theorem 1.6, and the remarks preceding Theorem 2.1 in Tarski [1955], or else Chang-Keisler [1973], Exercise 5.2.3), $\mathbf{S}(\{\mathfrak{A}_0, \ldots, \mathfrak{A}_{n-1}\})$ is a universal class. Applying Theorem 1.1, we see that $\mathbf{S}(\{\mathfrak{A}_0, \ldots, \mathfrak{A}_{n-1}\})$ is the class of simple models of the variety that it generates. But of course this coincides with the variety generated by $\{\mathfrak{A}_0, \ldots, \mathfrak{A}_{n-1}\}$, which is K. □

In Chapter 3 we shall be interested in the lattice of varieties of relation algebras. In view of Theorem 1.1 we may consider instead the lattice $\mathcal{S}$ of universal classes of simple relation algebras. Let M and N be two elements of $\mathcal{S}$ with $\mathsf{M} \subseteq \mathsf{N}$. Thus, M and N are universal classes of simple relation algebras. The collection (or *interval*) of all elements of $\mathcal{S}$ between M and N is denoted by $[\mathsf{M}, \mathsf{N}]$. Thus, $[\mathsf{M}, \mathsf{N}]$ is the collection of all universal subclasses of N that include M. It is closed under union and intersections, and is therefore a sublattice of $\mathcal{S}$.

For another consequence of open formula reducibility, recall that with any finite algebraic structure $\mathfrak{A}$ of finite similarity type, we can easily correlate an existential sentence $\varphi_{\mathfrak{A}}$ in the language of $\mathfrak{A}$ so that $\mathfrak{A}$ is embeddable into a structure $\mathfrak{B}$ (of the same similarity type) iff $\varphi_{\mathfrak{A}}$ is true of $\mathfrak{B}$ (see Tarski [1954]). Hence, the non-embeddability of $\mathfrak{A}$ is expressible by a universal sentence (equivalent to $\neg\varphi_{\mathfrak{A}}$). Combining this with open formula reducibility, we obtain the following:

LEMMA 1.3. *For a simple, finite relation algebra* $\mathfrak{A}$*, the non-embeddability of* $\mathfrak{A}$ *into simple relation algebras is expressible by an equation* $\varepsilon_{\mathfrak{A}}$*, i.e., a simple* $\mathfrak{B}$ *will not embed* $\mathfrak{A}$ *iff* $\varepsilon_{\mathfrak{A}}$ *is universally valid in* $\mathfrak{B}$*.*

We recall a special case of the notion of a semantic interpretation of one class of algebras into another (see Rabin [1965]). To semantically *define* or *interpret* semigroups into relation algebras, we need formulas $\varphi(v_0)$ and $\vartheta(v_0, v_1, v_2)$ (of the language $\mathcal{L}_r$ of relation algebras) that respectively define, in each relation algebra $\mathfrak{A}$, a subset $\varphi^{\mathfrak{A}}$ that is the universe of the semigroup, and a binary operation $\vartheta^{\mathfrak{A}}$ under which $\varphi^{\mathfrak{A}}$ is closed and which is associative on elements of $\varphi^{\mathfrak{A}}$. If φ and ϑ have additional parameter variables $w = (w_0, \ldots, w_{m-1})$, we may also specify a formula $\psi(w)$ giving conditions that the parameters are required to satisfy. In our applications, we shall only be concerned with open formulas. Thus, φ and ψ will always be open formulas, and ϑ will always be a term definition, i.e., an equation of the form $v_2 = \tau(v_0, v_1, w)$. Because we shall not have occasion to employ the word "interpretation" to discuss more general notions, we shall use it to denote the restricted notion we have just described.

DEFINITION 1.4. (i) An *interpretation* of semigroups into relation algebras is a triple

$$\langle \varphi(v_0, w)\,,\ \tau(v_0, v_1, w)\,,\ \psi(w)\rangle$$

(where φ and ψ are open formulas, and τ a term, of $\mathcal{L}_r$) such that, for every relation algebra $\mathfrak{A}$ and sequence a from $\mathfrak{A}$ satisfying ψ, the structure

$$\mathfrak{S}_{\mathfrak{A},a} = \langle \varphi^{\mathfrak{A}}[v_0, a]\,,\ \tau^{\mathfrak{A}}[v_0, v_1, a]\rangle$$

is a semigroup (in more detail, $\varphi^{\mathfrak{A}}$ is non-empty and closed under $\tau^{\mathfrak{A}}$, and $\tau^{\mathfrak{A}}$ restricted to elements of $\varphi^{\mathfrak{A}}$ is associative). When φ and τ have no parameter variables, we may drop ψ and treat $\langle \varphi\,,\ \tau\rangle$ as an interpretation.

(ii) An interpretation of semigroups into relation algebras *interprets finite semigroups into* K, a class of relation algebras, if, for every n in ω there is an $\mathfrak{A}$ in K and an a from $\mathfrak{A}$ satisfying the condition on the parameters (if any) such that $\mathfrak{S}(n)$ is embeddable into $\mathfrak{S}_{\mathfrak{A},a}$. □

REMARKS. Because we are only concerned with open formulas — and these are preserved under the formation of subalgebras — we do not have to require in (ii) that every finite semigroup is realizable as a defined semigroup. Rather, it suffices to require that each $\mathfrak{S}(n)$ is embeddable into some defined semigroup. Indeed, by the well-known Cayley representation, each finite semigroup is embeddable into $\mathfrak{S}(n)$ for some n in ω. (We may need to adjoin an identity element to the semigroup to insure that the embedding is one-one.) Thus, an open formula is universally valid in all finite semigroups iff it is universally valid in $\mathfrak{S}(n)$ for each n.

When the meaning is clear, we shall drop the reference to a in the notation $\mathfrak{S}_{\mathfrak{A},a}$ and write simply $\mathfrak{S}_{\mathfrak{A}}$. □

A set Φ of formulas is said to be *decidable*, or *recursive*, if its set of Gödel numbers (in some arithmetization of the language) is recursive, and *undecidable*, or *non-recursive*, otherwise. Two sets of formulas, Φ and Ψ, are *recursively inseparable* if there is no recursive set Θ of formulas such that $\Phi \subseteq \Theta$ and $\Psi \cap \Theta = \varnothing$. It will be very helpful for us to introduce a more specialized notion of recursive inseparability. We shall say that a single set Ψ of equations in the language $\mathcal{L}_r$ of relation algebras is *recursively inseparable* if $\mathcal{E}q(\mathsf{RA})$ and $\sim\Psi$ are recursively inseparable sets. (Of course $\sim\Psi$, the set of equations of $\mathcal{L}_r$ that are not in Ψ, is just the set of equations *falsifiable* in models of Ψ, i.e., the set of equations whose negation is satisfiable in some model of Ψ.) Equivalently, Ψ is recursively inseparable if there is no recursive set Θ of equations such that $\mathcal{E}q(\mathsf{RA}) \subseteq \Theta \subseteq \Psi$. In particular, if Ψ is a recursively inseparable equational theory of relation algebras, then every equational theory of relation algebras included in Ψ is undecidable. In other words, if K is a class of relation algebras with a recursively inseparable equational theory, and if L is any class of relation algebras such that $\mathsf{K} \subseteq \mathsf{L}$, then L must have an undecidable equational theory.

LEMMA 1.5. *Let* K *be a class of relation algebras. If there is an interpretation of semigroups into relation algebras that interprets finite semigroups into* $\mathbf{Si}(\mathsf{K})$, *then* $\mathcal{E}q(\mathsf{K})$ *is recursively inseparable.*

PROOF. Let $\mathcal{L}$ be the language of semigroups (with a binary operation symbol $\circ$), $\mathcal{L}_r$ the language of relation algebras, and

$$\langle \varphi(v_0, w)\,,\ \tau(v_0, v_1, w)\,,\ \psi(w)\rangle$$

an interpretation of semigroups into relation algebras, where w is an m-termed sequence of parameter variables. Without loss of generality we may assume that the distinct variables of $\mathcal{L}_r$ are $w_0, \ldots, w_{m-1}$ and $v_0, v_1, v_2, \ldots$ (thus, $w_i \neq v_j$ for every $i < m$ and $j < \omega$), and that the variables of $\mathcal{L}$ are $v_0, v_1, v_2, \ldots$ (thus, $w_0, \ldots, w_{m-1}$ do not occur in the vocabulary of $\mathcal{L}$).

We define a recursive mapping f from the terms of $\mathcal{L}$ to terms of $\mathcal{L}_r$ by stipulating that

$$f(v_i) = v_i \qquad \text{and} \qquad f(\sigma_0 \circ \sigma_1) = \tau(f(\sigma_0), f(\sigma_1), w)$$

for $i < \omega$ and terms σ_0, σ_1 of $\mathcal{L}$. We can extend f to a recursive mapping taking open formulas of $\mathcal{L}$ to open formulas of $\mathcal{L}_r$ by stipulating that $f(\vartheta)$ is obtained from ϑ by replacing each equation $\sigma_0 = \sigma_1$ in ϑ with the equation $f(\sigma_0) = f(\sigma_1)$. Using a simple and well-known argument, one establishes the following:

(1) For every relation algebra $\mathfrak{A}$ and sequence a from $\mathfrak{A}$ satisfying ψ, and for each open formula $\vartheta(v_0, \ldots v_{n-1})$ of $\mathcal{L}$ and each n-termed sequence b from $\varphi^{\mathfrak{A}}$, we have

$$\mathfrak{S}_{\mathfrak{A},a} \models \vartheta[b] \quad \text{iff} \quad \mathfrak{A} \models f(\vartheta)[b\hat{\ }a]\,.$$

The proof of (1) proceeds by induction on formulas. One begins by showing (by induction on terms) that

$$\sigma^{\mathfrak{S}_{\mathfrak{A}}}[b] = f(\sigma)^{\mathfrak{A}}[b\hat{\ }a]$$

for every term σ of $\mathcal{L}$. We leave the details to the reader.

Next, using f, we define another recursive mapping g from open formulas of $\mathcal{L}$ to open formulas of $\mathcal{L}_r$. For each open formula $\vartheta(v_0, \ldots v_{n-1})$ of $\mathcal{L}$ we stipulate that

$$g(\vartheta) = \big([\psi(w) \wedge \textstyle\bigwedge_{i<n}\varphi(v_i, w)] \rightarrow f(\vartheta)\big)\ .$$

With the help of (1) we easily obtain:

(2) If an open formula ϑ of $\mathcal{L}$ is universally valid in all semigroups, then $g(\vartheta)$ is universally valid in all relation algebras.

Indeed, suppose that ϑ is valid in all semigroups. Let $\mathfrak{A}$ be a relation algebra and b, a sequences from $\mathfrak{A}$ such that $b\hat{\ }a$ satisfies the hypotheses of $g(\vartheta)$. Thus, a satisfies ψ, and b is an n-termed sequence from $\varphi^{\mathfrak{A}}$. Since ϑ is valid in $\mathfrak{S}_{\mathfrak{A}}$ and b is from the universe of $\mathfrak{S}_{\mathfrak{A}}$, we obtain $\mathfrak{S}_{\mathfrak{A}} \models \vartheta[b]$. Therefore, $\mathfrak{A} \models f(\vartheta)[b\hat{\ }a]$, by (1). This shows that $b\hat{\ }a$ satisfies $g(\vartheta)$ in $\mathfrak{A}$.

In a similar fashion, we prove:

(3) If ϑ is an open formula of $\mathcal{L}$ that is falsifiable in a finite semigroup, then $g(\vartheta)$ is falsifiable in a simple algebra of K.

To see this, let ϑ be an open formula of $\mathcal{L}$ that is falsifiable in a finite semigroup. Since each finite semigroup is embeddable into $\mathfrak{S}(n)$ for some natural number n, since each $\mathfrak{S}(n)$ is embeddable into the defined semigroup of an algebra in $\mathsf{Si}(\mathsf{K})$, by 1.4(ii) and our assumption, and since ϑ is open, there must be a simple $\mathfrak{A}$ in K and a sequence a from $\mathfrak{A}$ satisfying ψ such that ϑ is falsifiable in $\mathfrak{S}_{\mathfrak{A},a}$ (see the remark after Definition 1.4). Let b be a sequence from $\mathfrak{S}_{\mathfrak{A}}$ that does not satisfy ϑ. Then $b\hat{}a$ does not satisfy $f(\vartheta)$ in $\mathfrak{A}$, by (1). Thus, $b\hat{}a$ satisfies the antecedent, but not the consequent, of $g(\vartheta)$ in $\mathfrak{A}$. This shows that $g(\vartheta)$ is falsifiable in $\mathfrak{A}$, and hence in $\mathsf{Si}(\mathsf{K})$.

Let Φ be the set of open formulas universally valid in all relation algebras, and Ψ the set of open formulas falsifiable in $\mathsf{Si}(\mathsf{K})$. We shall show that

(4) Φ and Ψ are recursively inseparable.

For contradiction, suppose that Δ is a recursive set such that

(5) $\Phi \subseteq \Delta$ and $\Psi \cap \Delta = \varnothing$.

Without loss of generality, we may assume that Δ consists just of open formulas of $\mathcal{L}_r$, since the set of such open formulas is recursive, and the intersection of two recursive sets is recursive. Obviously, $g^{-1}[\Delta]$ is a recursive set of open formulas of $\mathcal{L}$ (since g and its domain are both recursive, as is Δ). Moreover, by (5) we have

$$g^{-1}[\Phi] \subseteq g^{-1}[\Delta] \qquad \text{and} \qquad g^{-1}[\Psi] \cap g^{-1}[\Delta] = \varnothing\,.$$

Thus, $g^{-1}[\Delta]$ is a recursive separation of $g^{-1}[\Phi]$ and $g^{-1}[\Psi]$. But $g^{-1}[\Phi]$ includes the set Φ' of open formulas valid in all semigroups, by (2). Further, $g^{-1}[\Psi]$ includes the set Ψ' of open formulas falsifiable in some finite semigroup, by (3). Therefore, $g^{-1}[\Delta]$ is a recursive separation of Φ' and Ψ'. But Gurevich-Lewis [1984], p. 185, showed that Φ' and Ψ' are recursively inseparable.[1] (In fact, the respective subsets of conditional equations are recursively inseparable.) Thus, we have a contradiction, which proves (4).

Let h be the recursive function mapping each open formula ϑ of $\mathcal{L}_r$ to the equation ε_ϑ of $\mathcal{L}_r$ that is equivalent with ϑ in all simple relation algebras (see the remarks about open formula reducibility that precede Theorem 1.1). Consider the set Ξ of equations true in all relation algebras, and the set Θ of equations falsifiable in $\mathsf{Si}(\mathsf{K})$.

(6) Ξ and Θ are recursively inseparable.

Suppose, for contradiction, that Γ is a recursive set such that

$$\Theta \subseteq \Gamma \qquad \text{and} \qquad \Theta \cap \Gamma = \varnothing\,.$$

Without loss of generality, we may assume that Γ is a set of equations of $\mathcal{L}_r$. As in the proof of (4), $h^{-1}[\Gamma]$ is recursive, and

$$h^{-1}[\Xi] \subseteq h^{-1}[\Gamma] \qquad \text{and} \qquad h^{-1}[\Theta] \cap h^{-1}[\Gamma] = \varnothing\,.$$

[1] Gurevich has communicated to us that the essential steps in the argument that Φ' and Ψ' are recursively inseparable occur already in Gurevich [1966].

Each formula ϑ in Φ is equivalent with $h(\vartheta)$ in all simple relation algebras. Since ϑ is universally valid in all relation algebras, $h(\vartheta)$ is universally valid in all simple relation algebras. It now follows from semi-simplicity that $h(\vartheta)$ is valid in all relation algebras, i.e., $h(\vartheta)$ is in Ξ. This establishes the inclusion $\Phi \subseteq h^{-1}[\Xi]$. A similar argument shows that $\Psi \subseteq h^{-1}[\Theta]$. Thus, $h^{-1}[\Gamma]$ is a recursive separation of Φ and Ψ, in contradiction to (4). This proves (6).

Finally, the set Υ of equations falsifiable in K (but not necessarily in $\mathbf{Si}(\mathsf{K})$) includes Θ. Thus, it, too, must be recursively inseparable from Ξ (since any recursive separation of Ξ and Υ would also be a recursive separation of Ξ and Θ). This completes the proof of the lemma. □

We conclude this chapter by establishing a general algebraic fact concerning the inheritability of local finiteness. Let us call a class K of algebras *uniformly locally finite* if, for each $k \in \omega$, there is a natural number n_k such that every k-generated algebra in $\mathbf{S}(\mathsf{K})$ has cardinality at most n_k.

LEMMA 1.6. *Let* K *be a class of algebras of finite similarity type. If* K *is uniformly locally finite, then* $\mathbf{HSP}(\mathsf{K})$ *is locally finite.*

PROOF. If every algebra in K is trivial, then the same is true of $\mathbf{HSP}(\mathsf{K})$, and we immediately obtain the desired conclusion. Suppose that K has a non-trivial algebra and satisfies the hypothesis of the lemma. Fix an arbitrary natural number $k > 0$. We must show that every k-generated algebra in $\mathbf{HSP}(\mathsf{K})$ is finite. By assumption there is a finite bound n_k on the sizes of the k-generated algebras in $\mathbf{S}(\mathsf{K})$. Since for a finite similarity type there are only finitely many pairwise non-isomorphic algebras of cardinality at most n_k, the class $\mathbf{S}(\mathsf{K})$ has, up to isomorphic copies, only a finite number of k-generated algebras. Choose one such algebra for each isomorphism type, and let $\mathfrak{B}$ be their direct product (in some order).

(1) $\mathfrak{B}$ and K have the same k-variable equational theory, i.e., an equation ε with variables among $v_0, \ldots, v_{k-1}$ holds in $\mathfrak{B}$ iff it holds in K.

Indeed,

$$\begin{aligned} \varepsilon \text{ holds in } \mathfrak{B} \quad &\text{iff} \quad \varepsilon \text{ holds in each factor of } \mathfrak{B} \\ &\text{iff} \quad \varepsilon \text{ holds in all } k\text{-generated algebras in } \mathbf{S}(\mathsf{K}) \\ &\text{iff} \quad \varepsilon \text{ holds in } \mathsf{K}. \end{aligned}$$

The first equivalence holds because $\mathfrak{B}$ is a direct product and ε an equation, the second because of the choice of the factors of $\mathfrak{B}$, and the third because ε is an equation with only k variables.

Because K contains a non-trivial algebra and $k > 0$, the K-free algebra on k generators, $\mathfrak{F}_k\mathsf{K}$, and the $\mathfrak{B}$-free algebra on k generators, $\mathfrak{F}_k\mathfrak{B}$, exist. It follows from (1) that these two algebras are isomorphic. (Recall that the free algebra on k generators over a class L of algebras is isomorphic to the quotient of the algebra of k-variable terms modulo the congruence relation of semantic equivalence between these terms in L; see, e.g., Henkin-Monk-Tarski [1971], Theorem 0.4.61.) But $\mathfrak{F}_k\mathfrak{B}$ is isomorphic to a subalgebra of ${}^{^kB}\mathfrak{B}$ (see, e.g., *op. cit.*, Theorem 0.4.50(i) and Definition 0.4.1), so it is finite. We conclude that $\mathfrak{F}_k\mathsf{K}$ is finite. Finally, every

k-generated algebra in $\mathbf{HSP}(\mathsf{K})$ is a homomorphic image of $\mathfrak{F}_k\mathsf{K}$ (see, e.g., *op. cit.*, Theorems 0.4.26(ii) and 0.4.24) and is therefore finite. □

A locally finite quasi-variety K is always uniformly locally finite: all of the k-generated algebras in K are homomorphic images of, and hence have cardinality at most that of, $\mathfrak{F}_k\mathsf{K}$. It turns out that local finiteness implies uniform local finiteness for a much more general category of algebraic classes than quasi-varieties.

LEMMA 1.7. *Let* K *be a class of algebras that is closed under the formation of ultraproducts. If* K *is locally finite, then it is uniformly so.*

PROOF. Notice, first of all, that $\mathbf{S}(\mathsf{K})$ is closed under ultraproducts, since K is. (An ultraproduct of subalgebras of algebras $\mathfrak{C}_n$ is a subalgebra of the ultraproduct of the $\mathfrak{C}_n$.) We now argue by contraposition. Suppose that K is not uniformly locally finite. Then there is a $k \in \omega$ such that, for each $n \in \omega$, there is a k-generated algebra $\mathfrak{A}_n$ in $\mathbf{S}(\mathsf{K})$ of cardinality at least n. Let $a_{n,0}, \ldots, a_{n,k-1}$ be a collection generators of $\mathfrak{A}_n$. Fix a non-principal ultrafilter I on ω, and let $\mathfrak{B}$ be the ultraproduct of $\langle \mathfrak{A}_n : n \in \omega \rangle$ modulo I. Then $\mathfrak{B}$ is in $\mathbf{S}(\mathsf{K})$ by our initial remark. Set $b_i = \langle a_{n,i} : n \in \omega \rangle$ for each $i < k$. If $\{b_0, \ldots, b_{k-1}\}$ generates a finite subalgebra of $\mathfrak{B}$, say of cardinality m, then this subalgebra can be completely described by a set Φ of first-order formulas $\varphi(v_0, \ldots, v_{k-1})$ that are satisfied by $\langle b_0, \ldots, b_{k-1} \rangle$ in $\mathfrak{B}$. It follows from the Fundamental Theorem of Ultraproducts (see Chang-Keisler [1973], Theorem 4.1.9) that $\langle a_{n,0}, \ldots, a_{n,k-1} \rangle$ must satisfy Φ in $\mathfrak{A}_n$ for infinitely many n. Therefore, these algebras $\mathfrak{A}_n$ all have cardinality m. But this contradicts the assumption that the cardinality of $\mathfrak{A}_n$ is at least n for each n. We conclude that $\{b_0, \ldots, b_{k-1}\}$ must generate an infinite subalgebra of $\mathfrak{B}$. Since $\mathfrak{B}$ is in $\mathbf{S}(\mathsf{K})$, so is this subalgebra. Therefore, K is not locally finite. □

THEOREM 1.8. *Let* K *be a class of algebras of finite similarity type, and suppose that* K *is closed under the formation of ultraproducts (e.g.,* K *is an elementary class). If* K *is locally finite, then so is* $\mathbf{HSP}(\mathsf{K})$.

PROOF. If K is locally finite, then it is uniformly so, by Lemma 1.7. Therefore, the desired conclusion follows at once from Lemma 1.6. □

It is not difficult to construct examples which show that the restriction in Lemma 1.6 and Theorem 1.8 to classes of algebras of finite similarity type is necessary.

CHAPTER II

UNDECIDABILITY

Our first theorem in this chapter gives a simple and useful criterion that can be easily applied to establish the undecidability of the equational theory of a class of relation algebras.

THEOREM 2.1. *Let* K *be a class of relation algebras such that, for each* $n \in \omega$, *there is a simple algebra in* K *with at least* n *subidentity elements. Then* $\mathcal{E}q(\mathsf{K})$ *is recursively inseparable.*

PROOF. By Lemma 1.5, it suffices to find an interpretation $\langle \varphi , \tau \rangle$ of semigroups into relation algebras that interprets finite semigroups into $\mathbf{Si}(\mathsf{K})$. Take φ to be the equation $v_0 = v_0$ and τ the term $v_0 ; v_1$. For each relation algebra $\mathfrak{A}$, the set $\varphi^{\mathfrak{A}}$ is just the universe of $\mathfrak{A}$, and $\tau^{\mathfrak{A}}$ is just the relative product operation. Thus, the semigroup $\mathfrak{S}_{\mathfrak{A}}$ is nothing other than the relative product reduct of $\mathfrak{A}$. We see from Definition 1.4(i) that $\langle \varphi , \tau \rangle$ is an interpretation of semigroups into relation algebras.

It remains to establish that, for any given n in ω, there is an $\mathfrak{A}$ in $\mathbf{Si}(\mathsf{K})$ such that the semigroup $\mathfrak{S}(n)$ is embeddable into $\mathfrak{S}_{\mathfrak{A}}$. Fix n, and let $\mathfrak{A}$ be a simple algebra in K with at least 2^n subidentity elements. Easy, well-known, Boolean algebraic considerations show that $\mathfrak{A}$ must have at least n non-zero, pairwise disjoint, subidentity elements, say $x_0, \dots, x_{n-1}$. It follows from the laws of relation algebra that

$$x_i ; x_j = x_i \cdot x_j = \begin{cases} 0 & \text{if } i \neq j \\ x_i & \text{if } i = j \end{cases}$$

(see, e.g., Chin-Tarski [1951], Corollary 3.12). Since $\mathfrak{A}$ is simple, we conclude without difficulty that

$$(1) \qquad (x_i ; 1 ; x_j) ; (x_k ; 1 ; x_\ell) = \begin{cases} x_i ; 1 ; x_\ell & \text{if } j = k \\ 0 & \text{otherwise} \end{cases} ,$$

$$(2) \qquad (x_i ; 1 ; x_j) \cdot (x_k ; 1 ; x_\ell) = \begin{cases} x_i ; 1 ; x_j & \text{if } i = k \text{ and } j = \ell \\ 0 & \text{otherwise} \end{cases} .$$

Define a mapping g from $Sb\,(n \times n)$ into A by:

$$(3) \qquad g(R) = \sum \{x_i ; 1 ; x_j : (i,j) \in R\}$$

for each R in $Sb\,(n \times n)$. It is straightforward to check that g is one-one. Indeed, by applying (2) and (3), we get the equivalence of the following three assertions for

$R, S \subseteq n \times n$:

$$\begin{aligned} & g(R) = g(S)\,; \\ & x_i\,;1\,;x_j \le g(R) \quad \text{iff} \quad x_i\,;1\,;x_j \le g(S) \quad , \quad \text{for all} \quad i,j < n\,; \\ & (i,j) \in R \quad \text{iff} \quad (i,j) \in S \quad , \quad \text{for all} \quad i,j < n\,. \end{aligned}$$

We verify that g is a homomorphism:

$$\begin{aligned} g(R|S) &= \textstyle\sum\{x_i\,;1\,;x_j : (i,j) \in R|S\} && \text{by (3),} \\ &= \textstyle\sum\{x_i\,;1\,;x_j : \text{ for some } k,\ (i,k) \in R \text{ and } (k,j) \in S\}\,, && \text{by definition of } |\,, \\ &= (\textstyle\sum\{x_i\,;1\,;x_k : (i,k) \in R\})\,;(\textstyle\sum\{x_\ell\,;1\,;x_j : (\ell,j) \in S\}) && \text{by (1) and distributivity,} \\ &= g(R)\,;g(S) && \text{by (3).} \end{aligned}$$

This completes the proof. □

REMARKS. One can actually show that the function g in the previous proof preserves converses as well. We refer to the numbered items from the proof and (continuing with that numeration) we recall the following relation algebraic laws (see, e.g., Chin-Tarski [1951], Theorems 3.5 and 1.7):

(4) $\quad y^{\smile} \le 1'$ iff $y \le 1'$, and $1^{\smile} = 1\,.$

For every binary relation R on n we have:

$$\begin{aligned} g(R^{-1}) &= \textstyle\sum\{x_i\,;1\,;x_j : (i,j) \in R^{-1}\} && \text{by (3),} \\ &= \textstyle\sum\{x_i\,;1\,;x_j : (j,i) \in R\}\,, && \text{by definition of } ^{-1}\,, \\ &= \textstyle\sum\{(x_j\,;1\,;x_i)^{\smile} : (j,i) \in R\} && \text{by (4),} \\ &= (\textstyle\sum\{x_j\,;1\,;x_i : (j,i) \in R\})^{\smile} && \text{by the distributivity of } ^{\smile}\,, \\ &= g(R)^{\smile} && \text{by (3).} \end{aligned}$$

Thus, g embeds $\langle Sb\,(n \times n)\,,\,|\,,\,^{-1}\rangle$ into $\langle A\,,\,;\,,\,^{\smile}\rangle$.

To understand the intuition behind the proof of the above theorem, imagine that $\mathfrak{A}$ is the full set relation algebra on a set U of cardinality at least n. Without loss of generality we may assume that $n \subseteq U$. For each $i < n$, let x_i be the singleton subidentity element $\{(i,i)\}$. Thus, for $i,j < n$ we have $x_i\,;1\,;x_j = \{(i,j)\}$. It follows from (3) that $g(R) = R$ for each R in $Sb\,(n \times n)$, i.e., g is the identity map on $Sb\,(n \times n)$.

Now for any set Γ of n non-zero, pairwise disjoint subidentity elements of $\mathfrak{A}$, the sequence $\langle y\,;1\,;z : y,z \in \Gamma\rangle$ behaves the same with respect to ; and $\cdot$ as the sequence $\langle\{(i,j)\} : i,j < n\rangle$ behaves with respect to $|$ and $\cap$. That is, without the help of 1', the relation algebra cannot distinguish in $y\,;1\,;z$ how many elements y and z have. Thus, the mapping g defined in (3) *must* behave analogously to

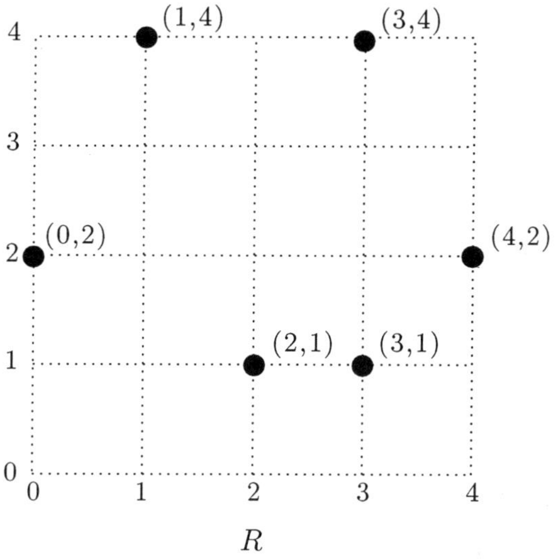

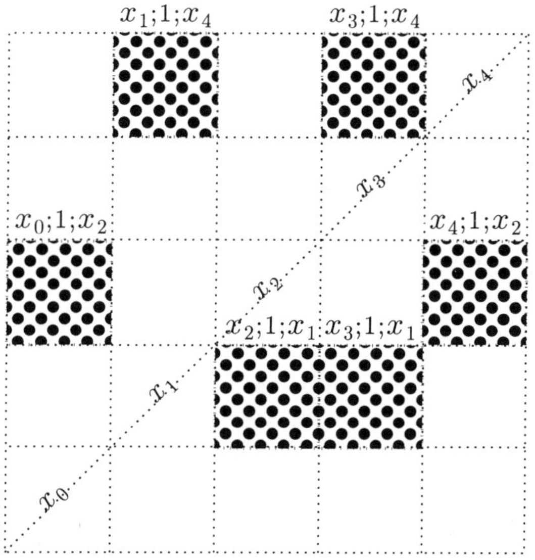

FIGURE 2.1. A relation $R = \{(0,2),(1,4),(2,1),(3,1),(3,4),(4,2)\}$ and its image $g(R)$.

the identity mapping on $Sb\,(n \times n)$, regardless of which sequence $\langle x_i : i < n\rangle$ of non-zero, pairwise disjoint, subidentity elements we use. (See Figure 2.1.)

The notion of a semi-associative relation algebra was introduced by Maddux. A set of axioms for these algebras is obtained by replacing the associative law for ; with a special case of that law:

$$(x\,;y)\,;1 = x\,;(y\,;1)\,.$$

Maddux [1991], Theorem 25, proved that every instance of a generalized associative law in which the symbol 1 occurs is true in all semi-associative relation algebras. For example,

$$(x ; 1) ; (y ; z) = [(x ; (1 ; y)] ; z$$

holds in all semi-associative relation algebras. Using this theorem, it is not difficult to see that Theorem 2.1 (and its proof) remain valid when K is taken to be a class of semi-associative relation algebras.

We shall show later that Theorem 2.1 is also a direct consequence of our principal undecidability result, Corollary 2.9. □

To prove the second undecidability theorem, we adapt to relation algebras the notion of an n-frame that was first formulated for complemented modular lattices in von Neumann [1960]. Simplifications and generalizations of this notion, with applications to decision problems, were given in Lipshitz [1974], Freese [1980], and Urquhart [1984], [1993]. First, we give three very weak frame axioms (substantially weaker than those in *op. cit.*). Then we show that, in the context of the theory of relation algebras, we can derive from these axioms the associativity of the (multiplicative) binary operation correlated with any given frame. Our presentation will be self-contained: the reader need not be familiar with the notion of frame or with any of the cited papers.

DEFINITION 2.2. Let $\mathfrak{A}$ be a relation algebra and $n \geq 4$. An n-frame in $\mathfrak{A}$ is a pair of sequences $\langle a_i : i < n \rangle$, $\langle c_{ij} : i, j < n, i \neq j \rangle$ satisfying the following conditions for all distinct $i, j, k, \ell < n$:

(i) $a_i^{\smile} = a_i$,
(ii) $(a_i ; a_j ; a_j ; a_\ell) \cdot (a_i ; a_k ; a_k ; a_\ell) \leq a_i ; a_\ell$,
(iii) $(c_{ij} ; c_{jk}) \cdot (a_i ; a_k) = c_{ik}$. □

REMARKS. In contrast to the definition, e.g., in Urquhart [1993], p. 9, we have not required that the elements a_i be transitive and commute with one another (i.e., $a_i ; a_i = a_i$ and $a_i ; a_j = a_j ; a_i$), nor that these elements together generate a Boolean algebra (with $;$ and $\cdot$ as the join and meet operations) of cardinality 2^n in which they are the atoms. (Lipshitz calls this last condition *independence*.) Instead, we have stipulated in condition (i) that the elements a_i are symmetric, and in condition (ii) that they satisfy a very weak special case of independence. We also have dropped the requirements of Lipshitz and Urquhart that

$$c_{ij} = c_{ji} \quad , \quad c_{ij} ; a_j = a_i ; a_j \quad , \quad c_{ij} \cdot a_j = c_{k\ell} \cdot a_\ell \quad ,$$

and that $c_{ij} \cdot a_j$ is an identity element (with respect to $;$) for the elements a_i and c_{ij}. However, we have retained the key condition (iii).

The symmetry condition (i) is not essential, but it does permit us to introduce some simplifications. As can be seen from the proof of Lemma 2.4(ii), we could replace conditions (i) and (ii) by the following two conditions:

$$(a_i ; a_j ; a_j ; a_\ell) \cdot (a_i ; a_k ; a_k{}^{\smile} ; a_\ell{}^{\smile}) \leq a_i ; a_\ell ,$$

$$(a_i ; a_j ; a_j ; a_\ell) \cdot (a_i{}^{\smile} ; a_k{}^{\smile} ; a_k ; a_\ell) \leq a_i ; a_\ell .$$

In our definition we have specified that $n \geq 4$ because of condition (ii). It is possible to reformulate this condition so as to allow also the notion of a 3-frame, but in this case the proof of Lemma 2.4(iii)–(v) does not go through. When $n < 3$, condition (iii) — which is the key condition — is vacuously satisfied. Thus, the notions of a 2-frame or a 1-frame are without interest and have been excluded. □

DEFINITION 2.3. Let $\langle a_i : i < n\rangle$, $\langle c_{ij} : i, j < n, i \neq j\rangle$ be an n-frame ($n \geq 4$) in a relation algebra $\mathfrak{A}$. For distinct $i, j, k < n$, set:

(i) $L_{ij} = \{x \in A : x \leq a_i ; a_j\}$;
(ii) $b \otimes d = (b ; d) \cdot (a_i ; a_k)$ for all $b \in L_{ij}$ and $d \in L_{jk}$;
(iii) $b \odot d = (b \otimes c_{12}) \otimes (c_{20} \otimes d)$ for all $b, d \in L_{01}$. □

REMARKS. Notice that c_{ij} is in L_{ij} and that $c_{ij} \otimes c_{jk} = c_{ik}$, by 2.2(iii) and 2.3(i),(ii).

Strictly speaking, the symbolic designations of the notions defined in 2.3 should bear a reference to $\mathfrak{A}$ and to the frame in terms of which they are defined, for example $L_{ij}(\mathfrak{A}, a, c)$. In practice, it should not cause any confusion to omit this reference. When discussing different frames simultaneously, it will usually suffice to refer to the relation algebra in which the frame is defined; in this case we will write, e.g., $L_{ij}(\mathfrak{A})$. In a similar vein, the symbol $\otimes$ should also bear a reference to the triple (i, j, k) of indices in terms of which it is defined. Again, we shall always omit such a reference.

Our definition of L_{ij} differs from that of Urquhart, who requires, e.g., that

$$x ; a_j = a_i ; a_j \quad , \quad x \cdot a_j = c_{ij} ; a_j \quad , \quad x ; c_{ij} ; a_j = c_{ij} ; a_j ; x = x\,.$$

In general, our set L_{ij} contains more elements than the set L_{ij} defined by Urquhart. Our definition of $\otimes$ is the one used by von Neumann and later authors. Our definition of $\odot$ is due to Freese [1980]. Urquhart [1984], p. 1061, points out that it "differs inessentially from that of von Neumann ... [and] is closely related to the well-known definition of multiplication [of points] on a line in a projective space due to von Staudt" (see the following example). □

We shall now give four examples of n-frames. The first two concern the notion of a frame that is used by the earlier authors. In fact, the first example is taken from von Neumann [1960], pp. 134–135. However, it is really a classical result of projective geometry that dates back to von Staudt [1857]. We shall concentrate on the case $n = 3$ in order to make clear the geometric motivation of the original frame notions and to explicate the meaning of the original frame axioms. (Recall our earlier remark that one could reformulate condition 2.2(ii) so as to admit 3-frames.) It is not necessary for the reader to understand the first two examples in order to follow our further development.

EXAMPLE. Consider a projective plane, say the real projective plane. The basic elements of our 3-frame will be points of the plane. Given two distinct points b and d, let $b ; d$ denote their join, i.e., the unique line through the two points. Dually, given two distinct lines, p and q, let $p \cdot q$ denote their unique point of intersection. The (original) independence condition says that a_0, a_1, and a_2 are pairwise distinct, non-collinear points. The original conditions on c_{ij} that we have dropped say that

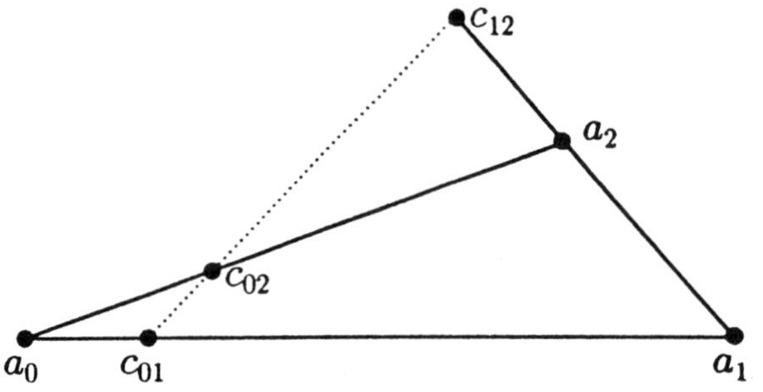

FIGURE 2.2

c_{ij} is collinear with, but distinct from a_i and a_j. Condition 2.2(iii) says that c_{01}, c_{12}, and c_{02} are collinear (see Figure 2.2).

The set L_{ij} (as originally defined) consists of the points collinear with a_i and a_j, but distinct from a_j. The point $b \otimes d$ is the point of intersection of the line through b and d with the line through a_i and a_k (see Figure 2.3)

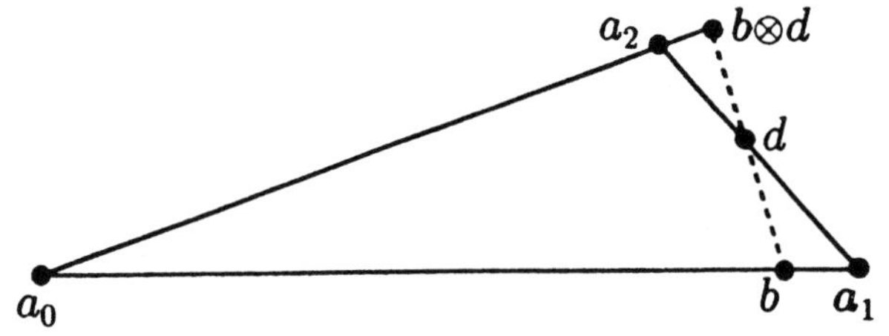

FIGURE 2.3

Finally, $b \odot d$ is the well-known multiplication of points on the line through a_0 and a_1 (see Figure 2.4), where a_0, c_{01} and a_1 play the roles of 0, 1 and ∞.

As is clear from the definition of $\odot$, we only need three coordinates, a_0, a_1, and a_2, to define multiplication. A fourth coordinate is needed to show axiomatically that the multiplication is associative (see the next lemma). Thus, we shall need only 4-frames in this work. $\square$

For von Neumann and Lipshitz, n-frames in complemented modular lattices of subspaces of vector spaces formed the key class of examples. Our second example describes these frames. In this and the next two examples we fix an integer $n \geq 3$.

EXAMPLE. Let $\mathfrak{V} = \langle V, +, -, \lambda \rangle_{\lambda \in F}$ be a vector space over some division ring F. The elements of the n-frame are κ-dimensional subspaces of $\mathfrak{V}$ (and thus elements of the group relation algebra $\mathfrak{Cm}(\mathfrak{V})$) for some fixed κ such that the dimension of $\mathfrak{V}$ is at least $n \cdot \kappa$. Indeed, the conditions of transitivity, commutativity, and independence given in the original frame definition essentially amount to the requirement that $a_0, \ldots, a_{n-1}$ be linearly independent subspaces of $\mathfrak{V}$. This means

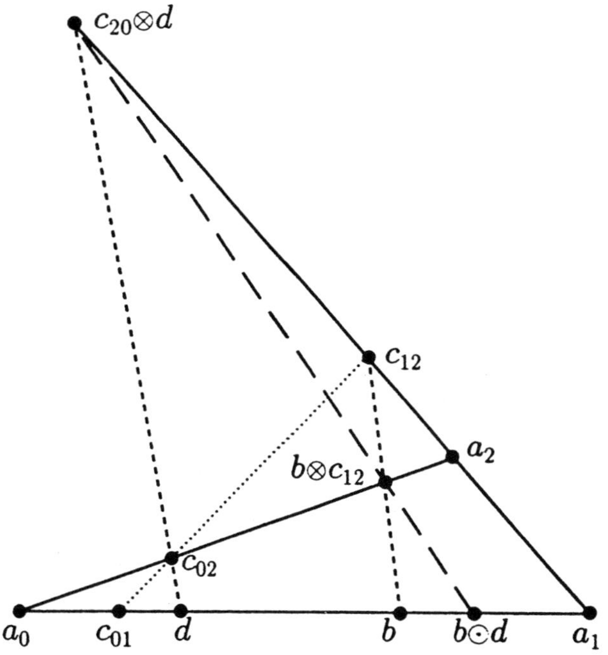

FIGURE 2.4. The definition of multiplication

that the sets of basis vectors for these subspaces are pairwise disjoint, and that their union forms a basis for the subspace generated by the union of $a_0, \ldots, a_{n-1}$. The element $a_i\,;a_j$ — the complex sum or relative product of the sets a_i and a_j that is formed in $\mathfrak{Cm}(\mathfrak{V})$ — is just the subspace of $\mathfrak{V}$ generated by a_i and a_j, i.e., it is the 2κ-dimensional plane passing through a_i and a_j. In analogy with the previous example, we could say that a_i and a_j are *2κ-coplanar*. Condition 2.2(iii) then says that c_{ij}, c_{jk}, and c_{ik} (for distinct $i, j, k < n$) are 2κ-coplanar. (See Figure 2.5 for the case when $n = 3$, $\mathfrak{V}$ is a 3-dimensional vector space over the real numbers, and $\kappa = 1$.)

The set L_{ij} (as originally defined) consists of the κ-dimensional planes that are 2κ-coplanar with a_i and a_j, but distinct from a_j. The point $b \otimes d$ is the κ-dimensional plane of intersection of the 2κ-dimensional plane through b and d with the 2κ-dimensional plane through a_i and a_k. In contrast to its interpretation under the original definition, under the current definition L_{ij} consists of all the subsets of $a_i\,;a_j$.

To be quite concrete we can take $\langle v_{i,\xi} : i < n,\ \xi < \kappa\rangle$ to be a system of linearly independent vectors of $\mathfrak{V}$. For each $i < n$, let a_i be the subspace of $\mathfrak{V}$ generated by $\{v_{i,\xi} : \xi < \kappa\}$, and for each relation $r \subseteq \kappa \times \kappa$ and each pair of distinct $i, j < n$, let r_{ij} be the subspace of $\mathfrak{V}$ generated by $\{v_{i,\xi} - v_{j,\eta} : (\xi, \eta) \in r\}$. For $r, s \subseteq \kappa \times \kappa$ and distinct $i, j, k < n$ we have

(1) $\quad r_{ij} \otimes s_{jk} = (r|s)_{ik}\,.$

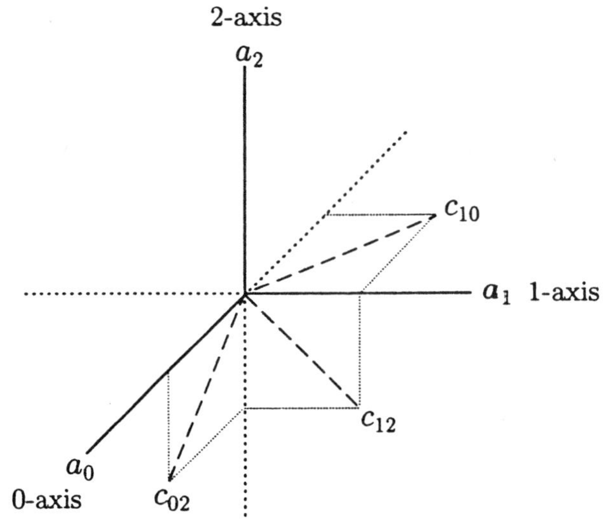

FIGURE 2.5. A 3-frame in the 3-dimensional vector space ${}^3\mathfrak{R}$ over the reals. The subspace a_i is the ith Cartesian axis and the subspace c_{ij} is the reflected diagonal in the (i, j)-plane.

(Since a calculation similar to the proof of (1) is carried out in the proof of step (4) of Lemma 2.7, we shall not stop to verify (1) now.) Taking c to be the identity relation on κ, and using (1) and the independence of the vectors $v_{i,\xi}$, it is not difficult to check that $\langle a_i : i < n \rangle$, $\langle c_{ij} : i, j < n, i \neq j \rangle$ is an n-frame in $\mathfrak{Cm}(\mathfrak{V})$. Using (1), a simple calculation also shows that

$$r_{01} \odot s_{01} = (r|s)_{01} .$$

From this it readily follows that $\langle L_{01}, \odot \rangle$ is a semigroup and the mapping $r \longmapsto r_{01}$ is an embedding of $\mathfrak{S}(\kappa)$ into $\langle L_{01}, \odot \rangle$. □

The next two examples are intended to give some insight into our more general approach to n-frames. In contrast to the previous two examples, their intuition is really relation algebraic in character, not geometric. We suggest that the reader check the (readily verifiable) assertions that we make (for example, about the validity of the frame axioms). Such checking will be helpful in clarifying the intuition of the frame axioms (for example, why 2.2(ii) gives us independence) and the relation algebraic intuition behind the definitions of the notions given in 2.3.

EXAMPLE. Let $\mathfrak{G} = \langle G, +, -, 0 \rangle$ be an arbitrary group. (Although we are using additive notation in this example, we are not assuming that the group is Abelian.) We construct an n-frame A, C in $\mathfrak{Cm}({}^n\mathfrak{G})$ by setting:

$$A_i = \{g \in {}^n\mathfrak{G} : g_p = 0 \text{ for } p \in n{\sim}\{i\}\},$$
$$C_{ij} = \{g \in {}^n\mathfrak{G} : g_j = -g_i \text{ and } g_p = 0 \text{ for } p \in n{\sim}\{i, j\}\}$$

for distinct $i, j < n$. (See Figure 2.5 for the case when $n = 3$ and $\mathfrak{G}$ is the additive group of the real numbers.) For example,

$$A_0 = \{(a, 0, \ldots, 0) : a \in G\},$$
$$C_{02} = \{(a, 0, -a, 0, \ldots, 0) : a \in G\}.$$

By definition of the operation ; in $\mathfrak{Cm}({}^n\mathfrak{G})$ we have

$$A_i \,;\, A_j = \{g \in {}^n\mathfrak{G} : g_p = 0 \text{ for } p \in n \sim \{i, j\}\}.$$

For example,

$$A_0 \,;\, A_2 = \{(a, 0, b, 0, \ldots, 0) : a, b \in G\}.$$

Thus, $A_i \,;\, A_j$ codes up the universal binary relation on G. By definition, L_{ij} is the collection of all subsets of $A_i \,;\, A_j$. Therefore it encodes the collection of all binary relations on G. However, the coding has a small twist to it in order to make computations work out properly. For each $R \subseteq G \times G$ and each pair of distinct $i, j < n$ set

$$R_{ij} = \{(0, \ldots, 0, f, 0, \ldots, 0, -g, 0, \ldots, 0) : (f, g) \in R\},$$

where f and $-g$ occur as the ith and j^{th} coordinates, respectively, of the n-tuple corresponding to (f, g). For example, if $R = \{(a, b), (a, c), (b, c)\}$, then

$$R_{02} = \{(a, 0, -b, 0, \ldots, 0), (a, 0, -c, 0, \ldots, 0), (b, 0, -c, 0, \ldots, 0)\}.$$

Notice that $C_{ij} = (I_G)_{ij}$, where I_G is the identity relation on G. One can show that for $R, S \subseteq G \times G$ and for distinct $i, j, k < n$ we have

$$R_{ij} \otimes S_{jk} = (R|S)_{ik}.$$

Therefore, a straightforward computation yields

$$R_{01} \odot S_{01} = (R|S)_{01}.$$

Moreover, for each $X \subseteq A_i \,;\, A_j$, setting

$$R_X = \{(g_i, -g_j) : g \in X\},$$

we have $(R_X)_{ij} = X$. From these observations it follows without great difficulty that $\langle L_{01}, \odot\rangle$ is a semigroup (in particular, L_{01} is closed under $\odot$) and that the correspondence $R \longmapsto R_{01}$ is an isomorphism of the semigroup $\mathfrak{S}(G)$ onto $\langle L_{01}, \odot\rangle$. (All these facts will be established in a more general setting in the proof of Lemma 2.7 and in the remark that follows it.) □

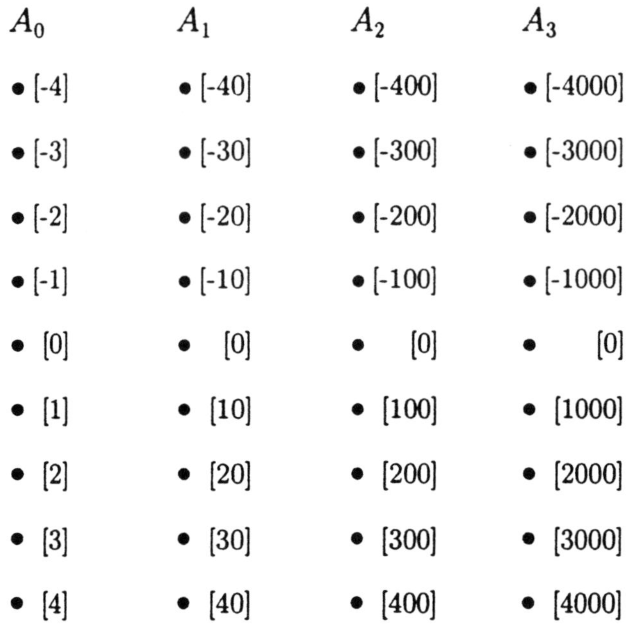

FIGURE 2.6. A 4-frame in $\mathfrak{Z}_m$ for some $m \geq 100,000$. The set C_{ij} can be thought of as a correspondence between the sets A_i and A_j.

EXAMPLE. Let m be a natural number at least as big as 10^{n+1} and let $\mathfrak{Z}_m$ be the additive group of integers modulo m. For each integer q, let $[q]$ denote the equivalence class of q in $\mathfrak{Z}_m$. (In this example we will use "p" and "q" as variables ranging over integers.) We construct an n-frame A, C in $\mathfrak{Cm}(\mathfrak{Z}_m)$ by setting

$$A_i = \{[10^i q] : -4 \leq q \leq 4\},$$
$$C_{ij} = \{[10^i q - 10^j q] : -4 \leq q \leq 4\}$$

for distinct $i, j < n$. For example,

$$A_1 = \{[-40], [-30], \ldots, [30], [40]\},$$
$$C_{12} = \{[-40 + 400], [-30 + 300], \ldots, [30 - 300], [40 - 400]\}.$$

(See Figure 2.6 for the case when $n = 4$.)

Using the definition of ; in $\mathfrak{Cm}(\mathfrak{Z}_m)$ we see that

$$A_i \,;\, A_j = \{[10^i p + 10^j q] : -4 \leq p, q \leq 4\}.$$

Thus, $A_i \,;\, A_j$ codes up the the universal binary relation on the 9-element set

$$U = \{-4, -3, \ldots, 3, 4\}.$$

Further, L_{ij} is the set of all subsets of $A_i \mathbin{;} A_j$, so it codes up the collection of all binary relations on U. Specifically, for each $R \subseteq U \times U$ and each pair of distinct $i, j < n$ set

$$R_{ij} = \{10^i p - 10^j q : (p, q) \in R\} .$$

As in the previous example, it is not difficult to verify that $C_{ij} = (I_U)_{ij}$ and that for $R, S \subseteq U \times U$ and for distinct $i, j, k < n$ we have

$$R_{ij} \otimes S_{jk} = (R|S)_{ik} ,$$

and therefore

$$R_{01} \odot S_{01} = (R|S)_{01} .$$

Moreover, for each $X \subseteq A_i \mathbin{;} A_j$, setting

$$R_X = \{(p, q) : [10^i p - 10^j q] \in X\} ,$$

we obtain $(R_X)_{ij} = X$. Consequently, $\langle L_{01} , \odot\rangle$ is a semigroup and that the correspondence $R \longmapsto R_{01}$ is an isomorphism of $\mathfrak{S}(U)$ onto $\langle L_{01} , \odot\rangle$.

By choosing bases larger than 10 and correspondingly larger m, it is possible to construct frames in $\mathfrak{Cm}(\mathfrak{Z}_m)$ that give us semigroups isomorphic to

$$\mathfrak{S}(\{-k, -k+1, \ldots, k-1, k\})$$

for larger (and in fact arbitrarily large) k. □

REMARKS. It is interesting to look at the earlier examples, say the second one, from the relation algebraic point of view. We can also think of $A_i \mathbin{;} A_j$ as coding up the universal binary relation on some fixed κ-dimensional subspace U of $\mathfrak{V}$. Then L_{ij} (as we have defined it) represents the collection of all binary relations on U, etc. Our particular definition of an n-frame (but not the original definition) would also allow us to take A_i to be the set $\{v_{i,\xi} : \xi < \kappa\}$ of κ linearly independent vectors, instead of the subspace generated by these vectors. In this case, $A_i \mathbin{;} A_j$ would represent the universal binary relation on κ and L_{ij} the collection of all binary relations on κ.

We shall show later (in step (4) of the proof of Lemma 2.7) that $\otimes$ and $\odot$ have the same type of interpretation in general as they do in the last two examples. □

Notice that conditions (i)–(iii) in 2.2 are expressed by equations. Thus, for a given n we can construct a conjunction $\psi(w)$ of equations, where w is a sequence of n^2 variables, such that the pair

$$a = \langle a_i : i < n\rangle \quad , \quad c = \langle c_{ij} : i, j < n , i \neq j\rangle$$

is an n-frame in $\mathfrak{A}$ iff $a^\frown c$ satisfies ψ in $\mathfrak{A}$.[2] Also, the set L_{01} is defined by an equation $\varphi(v_0, w)$, and the operations $\otimes$ and $\odot$ are defined by terms $\sigma(v_0, v_1, w)$

[2]Strictly speaking, for the notation $a^\frown c$ to make sense, c must be converted to a sequence (of length $n^2 - n$) indexed by ordinals. However, there is no problem in making such a conversion, so we will continue to use the concatenation notation.

and $\tau(v_0, v_1, w)$. Finally, notice that the only relation algebraic notions involved in the definition of an n-frame and in the definitions of L_{ij}, $\otimes$, and $\odot$ are ; , $\cdot$, and $\smile$.

In the next lemma we shall show that for $n \geq 4$ the triple $\langle \varphi , \tau , \psi \rangle$ just defined is an interpretation of semigroups into relation algebras. Essentially, this amounts to showing that the multiplication $\odot$ is an associative operation in each relation algebra in which a, c is an n-frame. (As is clear from Definition 2.3(iii), we only need three coordinates, a_0, a_1, and a_2, to define $\odot$. A fourth coordinate is needed to show that it is associative.) Afterwards, we shall show that, under certain restrictions on K (for example, that $\mathfrak{Cm}(\mathfrak{G})$ or $\mathfrak{Cf}(\mathfrak{G})$ is in K for some infinite group $\mathfrak{G}$), the triple $\langle \varphi , \tau , \psi \rangle$ interprets finite semigroups into $\mathbf{Si}(\mathsf{K})$. From this we shall draw several consequences about the undecidability of certain equational theories of relation algebras.

Parts (i) and (iii)–(v) of the lemma occur respectively as Lemma 5.1–Lemma 5.4 in Urquhart [1993]. Urquhart is following Lipshitz [1974], who in turn is essentially following von Neumann [1960]. Our proofs of (i) and (iii) are substantially different from those in *op. cit.*, since our definitions and axioms are weaker. However, our proofs of (iv) and (v) follow *op. cit.* almost verbatim. We shall need the following relation algebraic laws (and *only* these). They can all be found in Chin-Tarski [1951], or are immediate consequences of laws contained therein.

(L1) $(x\,;y)\,;z = x\,;(y\,;z)$,

(L2) $(x\,;y)^{\smile} = y^{\smile}\,;x^{\smile}$,

(L3) $x \cdot y \leq u \cdot v$ whenever $x \leq u$ and $y \leq v$,

(L4) $x\,;y \leq u\,;v$ whenever $x \leq u$ and $y \leq v$,

(L5) $(x\,;y) \cdot z \leq \big(x\,;[y \cdot (x^{\smile}\,;z)]\big) \cdot z$,

(L6) $(x\,;y) \cdot z \leq \big([x \cdot (z\,;y^{\smile})]\,;y\big) \cdot z$.

LEMMA 2.4. *Let $\langle a_i : i < n\rangle$, $\langle c_{ij} : i,j < n\,, i \neq j\rangle$ be an n-frame in a relation algebra $\mathfrak{A}$, with $n \geq 4$. Further, let $i, j, k, \ell < n$ be distinct.*

(i) *If $b \in L_{ij}$ and $d \in L_{jk}$, then $b \otimes d \in L_{ik}$.*
(ii) *If $b, d \in L_{01}$, then $b \odot d \in L_{01}$.*
(iii) *If $b \in L_{ij}$, $d \in L_{jk}$, and $f \in L_{k\ell}$, then*

$$(b \otimes d) \otimes f = b \otimes (d \otimes f)\,.$$

(iv) *If $b, d \in L_{01}$, then*

$$(b \otimes c_{12}) \otimes (c_{20} \otimes d) = (b \otimes c_{13}) \otimes (c_{30} \otimes d)\,.$$

(v) *For all $b, d, f \in L_{01}$, we have*

$$(b \odot d) \odot f = b \odot (d \odot f)\,.$$

PROOF. Part (i) follows immediately from 2.3(i),(ii), while part (ii) is a simple consequence of (i) and 2.3(iii). We turn to the proof of (iii), and we begin by showing that

(1) $(b \otimes d) \otimes f = [(b\,;d)\,;f] \cdot (a_i\,;a_\ell)\,.$

Notice that, by definition of $\otimes$, we have $b \otimes d \leq b\,; d$. Since $b \otimes d \in L_{ik}$, by part (i), we obtain from 2.3(ii) and the monotony laws (L3) and (L4) that

$$(b \otimes d) \otimes f = [(b \otimes d)\,;f] \cdot (a_i\,;a_\ell) \leq [(b\,;d)\,;f] \cdot (a_i\,;a_\ell)\,.$$

To establish the reverse inclusion, notice that $f^{\smile} \leq a_\ell\,;a_k$, by 2.2(i), (L2), and the assumption that $f \in L_{k\ell}$. Hence,

$$a_i\,;a_\ell\,;f^{\smile} \leq a_i\,;a_\ell\,;a_\ell\,;a_k\,,$$

by (L4). Also, since $b \in L_{ij}$ and $d \in L_{jk}$, we have

$$b\,;d \leq a_i\,;a_j\,;a_j\,;a_k\,,$$

by (L4) and 2.3(i). Combining these last two inequalities, and applying (L3) and 2.2(ii), we obtain

(2) $(b\,;d) \cdot (a_i\,;a_\ell\,;f^{\smile}) \leq a_i\,;a_k\,.$

Taking $b\,;d$ for x, f for y, and $a_i\,;a_\ell$ for z in (L6), we see that

$$[(b\,;d)\,;f] \cdot (a_i\,;a_\ell) \leq \big([(b\,;d) \cdot (a_i\,;a_\ell\,;f^{\smile})]\,;f\big) \cdot (a_i\,;a_\ell)\,.$$

Combining this with (2), and using (L3), (L4), we arrive at

$$[(b\,;d)\,;f] \cdot (a_i\,;a_\ell) \leq \big([(b\,;d) \cdot (a_i\,;a_k)]\,;f\big) \cdot (a_i\,;a_\ell)\,,$$

i.e.,

$$[(b\,;d)\,;f] \cdot (a_i\,;a_\ell) \leq (b \otimes d) \otimes f\,.$$

This proves (1).

In a similar fashion, we prove

(3) $b \otimes (d \otimes f) = [b\,;(d\,;f)] \cdot (a_i\,;a_\ell)\,.$

In view of (L1), we obtain (iii) from (1) and (3).

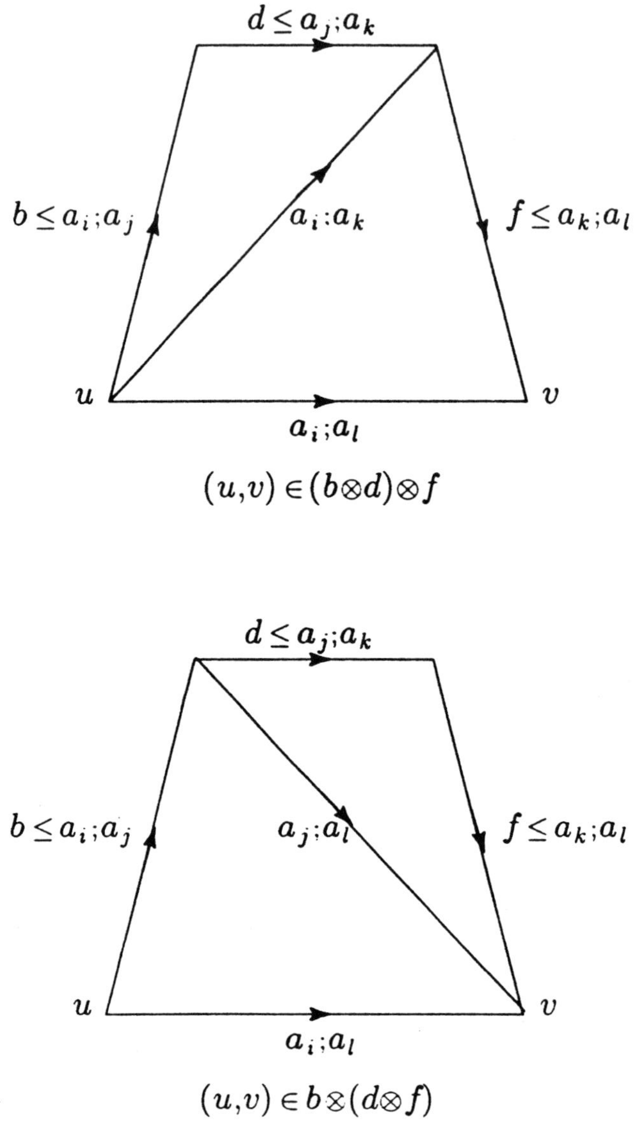

FIGURE 2.7. Illustration of a proof of 3.4(iii) for the case of a set relation algebra.

Next, we take up the proof of (iv). As was pointed out, 2.2(iii) and 2.3(ii) give us

(4) $c_{pr} \otimes c_{rq} = c_{pq}$ for all distinct $p, q, r < n$.

Consequently,

$$(b \otimes c_{12}) \otimes (c_{20} \otimes d) = [b \otimes (c_{13} \otimes c_{32})] \otimes (c_{20} \otimes d) \qquad \text{by (4)},$$

$$
\begin{aligned}
&= [(b \otimes c_{13}) \otimes c_{32}] \otimes (c_{20} \otimes d) && \text{by (iii)},\\
&= (b \otimes c_{13}) \otimes [c_{32} \otimes (c_{20} \otimes d)] && \text{by (iii),(i)},\\
&= (b \otimes c_{13}) \otimes [(c_{32} \otimes c_{20}) \otimes d] && \text{by (iii)},\\
&= (b \otimes c_{13}) \otimes (c_{30} \otimes d) && \text{by (4)}.
\end{aligned}
$$

Finally, we prove (v):

$$
\begin{aligned}
(b \odot d) \odot f &= \big([(b \otimes c_{12}) \otimes (c_{20} \otimes d)] \otimes c_{12}\big) \otimes (c_{20} \otimes f) && \text{by 2.3(iii)},\\
&= \big([(b \otimes c_{13}) \otimes (c_{30} \otimes d)] \otimes c_{12}\big) \otimes (c_{20} \otimes f) && \text{by (iv)},\\
&= \big((b \otimes c_{13}) \otimes [(c_{30} \otimes d) \otimes c_{12}]\big) \otimes (c_{20} \otimes f) && \text{by (iii),(i)},\\
&= (b \otimes c_{13}) \otimes \big([(c_{30} \otimes d) \otimes c_{12}] \otimes (c_{20} \otimes f)\big) && \text{by (iii),(i)},\\
&= (b \otimes c_{13}) \otimes \big([c_{30} \otimes (d \otimes c_{12})] \otimes (c_{20} \otimes f)\big) && \text{by (iii)},\\
&= (b \otimes c_{13}) \otimes \big(c_{30} \otimes [(d \otimes c_{12}) \otimes (c_{20} \otimes f)]\big) && \text{by (iii),(i)},\\
&= (b \otimes c_{12}) \otimes \big(c_{20} \otimes [(d \otimes c_{12}) \otimes (c_{20} \otimes f)]\big) && \text{by (iv),(i)},\\
&= b \odot (d \odot f) && \text{by 2.3(iii)}. \qquad \square
\end{aligned}
$$

COROLLARY 2.5. *If $n \geq 4$, then $\langle \varphi, \tau, \psi \rangle$ is an interpretation of semigroups into relation algebras.*

PROOF. Assume $n \geq 4$. Let $\mathfrak{A}$ be any relation algebra and

$$a = \langle a_i : i < n \rangle \quad , \qquad c = \langle c_{ij} : i, j < n, i \neq j \rangle$$

any sequences from $\mathfrak{A}$ such that $a\hat{\ }c$ satisfies ψ, i.e., such that a, c is an n-frame. The set $\varphi^{\mathfrak{A}}[v_0, a\hat{\ }c]$ is closed under the binary operation $\tau^{\mathfrak{A}}[v_0, v_1, a\hat{\ }c]$, by 2.3(iii) and 2.4(i), and this operation is associative on $\varphi^{\mathfrak{A}}$, by 2.4(v). In other words, $\mathfrak{S}_{\mathfrak{A}} = \langle \varphi^{\mathfrak{A}}, \tau^{\mathfrak{A}} \rangle$ is a semigroup. In view of Definition 1.4(i), this establishes the corollary. $\square$

Since the formula ψ and the term τ define the set L_{01} and the operation $\odot$ respectively, the semigroup $\mathfrak{S}_{\mathfrak{A}}$ referred to in the previous proof is just $\langle L_{01}, \odot \rangle$. We shall denote this semigroup by $\mathfrak{L}_{01}$ (or by $\mathfrak{L}_{01}(\mathfrak{A})$ when a reference to $\mathfrak{A}$ is needed), and we shall call it the *semigroup associated with the frame a, c in $\mathfrak{A}$*, or, for short, the *associated frame semigroup*.

As the previous lemma indicates, it will suffice for our purposes to consider just 4-frames. From now on we restrict our attention to such frames.

The following technical lemma is needed for the proof of Lemma 2.7.

LEMMA 2.6. *Let $\mathfrak{G} = \langle G, \circ, ^{-1}, e \rangle$ be an infinite group. Then there is a finite, cyclic subgroup H of $\mathfrak{G}$ such that, for each sequence $m = \langle m_i : i < 4 \rangle$ of natural numbers, there is a sequence $A = \langle A_i : i < 4 \rangle$ of subsets of G satisfying the following conditions:*

(i) $H \subseteq A_i$ *and* $m_i \leq |A_i| < \omega$ *for* $i < 4$;
(ii) $A_i = A_i^{-1} = A_i \circ H$ *for* $i < 4$;
(iii) *Whenever $b_i \in A_i^2$ for $i < 4$, and π is a permutation of 4 such that*

$$b_{\pi(0)} \circ b_{\pi(1)} \circ b_{\pi(2)} \circ b_{\pi(3)} \in H,$$

then $b_i \in H$ for each $i < 4$.

PROOF. Consider the equivalence relation $\equiv$ on G defined by the stipulation:

$$g \equiv h \qquad \text{iff} \qquad g^2 = h^2 .$$

If $\equiv$ has infinitely many equivalence classes, take X to be a set of representatives for the equivalence classes, and take H to be $\{e\}$. If $\equiv$ has only finitely many equivalence classes, then at least one such class is infinite (since G is infinite); take X to be some infinite equivalence class, and take H to be the subgroup of $\mathfrak{G}$ generated by $\hat{g}^2$ for some fixed $\hat{g} \in X$. (Recall that, in this case, $g^2 = h^2$ for all $g, h \in X$. Thus, H is independent of the element of X that we choose to define it.)

Before beginning the construction of $\langle A_i : i < 4\rangle$, we establish the following properties of X and H.

(1) X is an infinite subset of G, and H is a finite, cyclic subgroup of $\mathfrak{G}$.

(2) For every g in $(X \circ H) \cup (X^{-1} \circ H) \cup H$ and every h in H, we have $g \circ h = h \circ g$.

(3) For all finite $B, D \subseteq G$, there are infinitely many elements g in X such that $g \notin B \circ D \circ B$ and $g^2 \notin B \sim H$.

In the case when $\equiv$ has infinitely many equivalence classes, (1) and (2) are obvious, since $H = \{e\}$. For (3), observe that, by the choice of X, $g^2 \neq h^2$ for distinct g, h in X. Thus, both X and $\{g^2 : g \in X\}$ are infinite. Since both $B \circ D \circ B$ and B are finite, there must be infinitely many g in X such that $g \notin B \circ D \circ B$ and $g^2 \notin B \sim H$.

Assume now that $\equiv$ has finitely many equivalence classes. The set X is infinite, by definition. Each element of H has the form h^2 for some $h \in G$. (Indeed, if $x \in H$, then $x = (\hat{g}^2)^n$ for some integer n; set $h = \hat{g}^n$, and observe that $x = h^2$.) Thus, if H were infinite, then $\equiv$ would have infinitely many equivalence classes, in contradiction to the case that we are in. This proves (1). To prove (2), suppose that $g \in X$. Then $g^2 = \hat{g}^2$, by definition of X. Therefore, H, being generated by $\hat{g}^2$, is included in the subgroup generated by g. But this latter subgroup is cyclic, and hence Abelian. Thus, both g and g^{-1} commute with elements of H. This shows that the elements of $X \cup X^{-1}$ commute with elements of H. Clearly, then, elements of $(X \circ H) \cup (X^{-1} \circ H)$ must also commute with elements of H (since H is Abelian). We now turn to (3). For each $g \in X$, we have $g^2 \in H$ (since $g^2 = \hat{g}^2$), and therefore $g^2 \notin B \sim H$. Because X is infinite and $B \circ D \circ B$ is finite, $X \sim B \circ D \circ B$ must be infinite.

We now construct $\langle A_i : i < 4\rangle$ by induction on $p = m_0 + m_1 + m_2 + m_3$. Moreover, we shall construct this sequence so that

(4) $A_i \subseteq (X \circ H) \cup (X^{-1} \circ H) \cup H \quad$ for $i < 4$.

Thus, in view of (2), we will obtain the additional property:

(5) Every element of $\bigcup_{i<4} A_i$ commutes with every element of H.

When $p = 0$ (so that $m_i = 0$ for $i < 4$), we simply take $A_i = H$ for each $i < 4$. Conditions (i)–(iii) and (4) are trivial to verify, using the fact that H is a subgroup.

Now suppose that $p > 0$, and that the assertion holds for all quadruples with sum $p-1$. Because $p > 0$, some m_i is positive, say $m_0 > 0$. By the induction hypothesis, we can construct a sequence $\langle \bar{A}_i : i < 4 \rangle$ satisfying conditions (i)–(iii) and (4) for the quadruple $\langle m_0 - 1, m_1, m_2, m_3 \rangle$. Let

$$(6) \qquad B = \bigcup \{ \bar{A}_i^2 \circ \bar{A}_j^2 \circ \bar{A}_k^2 : 1 \le i, j, k \le 3 \text{ and } \{i, j, k\} = \{1, 2, 3\} \} .$$

Then B and $\bar{A}_0$ are finite subsets of G. By (3), there is a g such that

$$(7) \qquad g \in X \sim \bar{A}_0 \quad , \quad g \notin B \circ \bar{A}_0 \circ B \quad , \quad g^2 \notin B \sim H .$$

Set

$$(8) \qquad A_0 = \bar{A}_0 \cup (g \circ H) \cup (g^{-1} \circ H), \quad \text{and} \quad A_i = \bar{A}_i \quad \text{for } i = 1, 2, 3 .$$

Observe that condition (4) — and hence also (5) — holds, since $g \in X$. Also,

$$(9) \qquad \bar{A}_i \subseteq A_i \quad \text{for } i < 4 .$$

Furthermore, $H \subseteq A_0$ (since $H \subseteq \bar{A}_0$), and A_0 is finite (since $\bar{A}_0$ and H are finite). Because $g \notin \bar{A}_0$, we have

$$|A_0| \ge |\bar{A}_0| + 1 \ge (m_0 - 1) + 1 = m_0 .$$

This verifies condition (i). Condition (ii) is equally easy to verify. For A_1, A_2, and A_3, it holds by the induction hypotheses on $\bar{A}_1$, $\bar{A}_2$, and $\bar{A}_3$. For $\bar{A}_0$ we have:

$A_0^{-1} = \bar{A}_0^{-1} \cup (g \circ H)^{-1} \cup (g^{-1} \circ H)^{-1}$	by (8) and the distributive law for $^{-1}$ over $\cup$,
$= \bar{A}_0^{-1} \cup (H \circ g^{-1}) \cup (H \circ g)$	by (1),
$= \bar{A}_0 \cup (g^{-1} \circ H) \cup (g \circ H)$	by (5), and (ii) for $\bar{A}_0$,
$= A_0$	by (8);

$A_0 \circ H = (\bar{A}_0 \circ H) \cup (g \circ H^2) \cup (g^{-1} \circ H^2)$	by (8) and the distributive law for $\circ$ over $\cup$,
$= \bar{A}_0 \cup (g \circ H) \cup (g^{-1} \circ H)$	by (1), and (ii) for $\bar{A}_0$,
$= A_0$	by (8).

It remains to verify condition (iii). Using the distributive law for $\circ$ over $\cup$, (5), (8), and (ii) for $\bar{A}_0$, a straightforward computation shows that

$$(10) \quad A_0^2 = \bar{A}_0^2 \cup (g \circ \bar{A}_0) \cup (g^{-1} \circ \bar{A}_0) \cup (\bar{A}_0 \circ g) \cup (\bar{A}_0 \circ g^{-1}) \cup (g^2 \circ H) \cup (g^{-2} \circ H) .$$

For example, using (8) and the distributive law, we see that

$$g \circ H \circ \bar{A}_0 \quad , \quad \bar{A}_0 \circ g^{-1} \circ H \quad , \quad g^{-1} \circ H \circ g^{-1} \circ H \quad , \quad g \circ H \circ g^{-1} \circ H$$

are among the nine summands of A_0^2. By (5) and (ii) for $\bar{A}_0$, we have

$$\begin{aligned} g\circ H\circ \bar{A}_0 &= g\circ \bar{A}_0\circ H = g\circ \bar{A}_0\,,\\ \bar{A}_0\circ g^{-1}\circ H &= \bar{A}_0\circ H\circ g^{-1} = \bar{A}_0\circ g^{-1}\,,\\ g^{-1}\circ H\circ g^{-1}\circ H &= g^{-1}\circ g^{-1}\circ H^2 = g^{-2}\circ H\,. \end{aligned}$$

Since $H\subseteq \bar{A}_0$, we also have

$$g\circ H\circ g^{-1}\circ H = g\circ g^{-1}\circ H^2 = H^2 \subseteq \bar{A}_0^2\,.$$

We leave the remaining details of the proof of (10) to the reader.

Now let $b_i \in A_i^2$ for $i<4$. Suppose that π is a permutation of 4 such that

(11) $\quad b_{\pi(0)}\circ b_{\pi(1)}\circ b_{\pi(2)}\circ b_{\pi(3)} \in H\,.$

Clearly, $b_i^{-1}\in A_i^{-2}$. Applying conditions (ii) and (5), we obtain

$$\begin{aligned} H\circ b_{\pi(3)}^{-1}\circ b_{\pi(2)}^{-1}\circ b_{\pi(1)}^{-1} &\subseteq H\circ A_{\pi(3)}^{-2}\circ A_{\pi(2)}^{-2}\circ A_{\pi(1)}^{-2}\\ &= H\circ A_{\pi(3)}^{2}\circ A_{\pi(2)}^{2}\circ A_{\pi(1)}^{2}\\ &= A_{\pi(3)}^{2}\circ A_{\pi(2)}^{2}\circ A_{\pi(1)}^{2}\,. \end{aligned}$$

Since $b_{\pi(0)}$ is in $H\circ b_{\pi(3)}^{-1}\circ b_{\pi(2)}^{-1}\circ b_{\pi(1)}^{-1}$, by (11), we conclude that

$$b_{\pi(0)} \in A_{\pi(3)}^{2}\circ A_{\pi(2)}^{2}\circ A_{\pi(1)}^{2}\,.$$

In a similar fashion, we show that

$$\begin{aligned} b_{\pi(1)} &\in A_{\pi(0)}^{2}\circ A_{\pi(3)}^{2}\circ A_{\pi(2)}^{2}\,,\\ b_{\pi(2)} &\in A_{\pi(1)}^{2}\circ A_{\pi(0)}^{2}\circ A_{\pi(3)}^{2}\,,\\ b_{\pi(3)} &\in A_{\pi(2)}^{2}\circ A_{\pi(1)}^{2}\circ A_{\pi(0)}^{2}\,. \end{aligned}$$

Now $\pi(i)=0$ for some $i<4$, so for this i we have

$$\{\pi(j) : j<4\,,\ j\neq i\} = \{1,2,3\}\,.$$

In view of the definition (6) of B, and the fact that $b_0 = b_{\pi(i)}$, we get

(12) $\quad b_0 \in B\,.$

We now argue that

(13) $\quad b_0 \in \bar{A}_0^2\,.$

Since $b_0\in A_0^2$, by assumption, we see that b_0 must be in one of the summands on the right-hand side of the equation in (10). Suppose, for example, that $b_0\in g\circ \bar{A}_0$, say $b_0 = g \circ a_0$. Then $g = b_0 \circ a_0^{-1} \circ e$ (recall that e is the identity element), and this latter element is in $B \circ \bar{A}_0 \circ B$. (Indeed,

$$H = H^6 \subseteq A_i^2 \circ A_j^2 \circ A_k^2 \subseteq B\,,$$

so we get $e \in B$. Further, we have $a_0^{-1} \in \bar{A}_0^{-1} = \bar{A}_0$, by (ii) for $\bar{A}_0$, and $b_0 \in B$, by (12).) But $g \notin B \circ \bar{A}_0 \circ B$, by (7). Thus, we cannot have b_0 in $g \circ \bar{A}_0$. Similar arguments show that b_0 is not in $g^{-1} \circ \bar{A}_0$, nor in $\bar{A}_0 \circ g$, nor in $\bar{A}_0 \circ g^{-1}$.

Suppose, now, that b_0 is in $g^{-2} \circ H$, say $b_0 = g^{-2} \circ h$. Then $g^2 = h \circ b_0^{-1}$. Now

$$H \circ B^{-1} = H \circ B = B\,,$$

by (6), condition (ii), and (5). Therefore, $g^2 \in B$. But by (7), this implies that $g^2 \in H$. It follows that $g^{-2} \circ h$, i.e., b_0, is in H. Since $H = H^2 \subseteq \bar{A}_0^2$, we obtain (13). A similar argument applies if b_0 is in $g^2 \circ H$, and of course the case when $b_0 \in \bar{A}_0^2$ is trivial. This completes the proof of (13).

By (13), we have $b_0 \in \bar{A}_0^2$. Moreover, for $i = 1, 2, 3$ we have $b_i \in \bar{A}_i^2$, because $A_i = \bar{A}_i$. In view of (11) and the induction hypothesis for $\langle \bar{A}_i : i < 4 \rangle$ (as it applies to condition (iii)), we obtain $b_i \in H$ for each $i < 4$. This completes the verification of conditions (i)–(iii) and of (4). □

REMARKS. The statement and proof of Lemma 2.6 simplify considerably and become much more transparent if for $\mathfrak{G}$ we take the group $\mathfrak{Z} = \langle Z\,, +\,, -\,, 0 \rangle$ of integers. In this case the lemma can be reformulated as follows:

LEMMA 2.6 FOR $\mathfrak{Z}$. *For each sequence $m = \langle m_i : i < 4 \rangle$ of natural numbers there is a sequence $A = \langle A_i : i < 4 \rangle$ of subsets of Z satisfying the following conditions:*

(i) $0 \in A_i$ *and* $m_i \le |A_i| < \omega$ *for* $i < 4$;

(ii) $A_i = -A_i$ *for* $i < 4$;

(iii) *Whenever* $b_i \in A_i + A_i$ *for* $i < 4$, *and* $b_0 + b_1 + b_2 + b_3 = 0$, *then* $b_i = 0$ *for each* $i < 4$.

We see that (iii) asserts the linear independence of the sets $A_i + A_i$ in the sense that each sequence of elements b_i from these sets is linearly independent. To help the reader understand the intuition behind the proof of Lemma 2.6, we shall give the simplified proof of the above special case.

PROOF. We proceed by induction on $p = \sum_{i<4} m_i$. When $p = 0$, take $A_i = \{0\}$. Conditions (i)–(iii) are trivial to verify. Suppose now that $p > 0$ and that the construction is possible for all quadruples of natural numbers whose sum is less that p. By the induction hypothesis we can construct a sequence $\langle \bar{A}_i : i < 4 \rangle$ satisfying (i)–(iii) for $(m_0 - 1, m_1, m_2, m_3)$. Set

(1) $\quad B = (\bar{A}_1 + \bar{A}_1) + (\bar{A}_2 + \bar{A}_2) + (\bar{A}_3 + \bar{A}_3)\,.$

Observe that

(2) $\quad B = -B \quad , \quad 0 \in B \quad , \quad B \subseteq B + \bar{A}_0\,,$

by conditions (ii) and (i) for the $\bar{A}_i$. Choose an integer n that is greater than all numbers in $B + \bar{A}_0$. This is possible because $B + \bar{A}_0$ is finite. Set

$$A_0 = \bar{A}_0 \cup \{-n, n\} \qquad \text{and} \qquad A_i = \bar{A}_i \quad \text{for } 1 \leq i \leq 3 .$$

Using conditions (i) and (ii) for $\langle \bar{A}_i : i < 4\rangle$, it is simple to verify (i) and (ii) for $\langle A_i : i < 4\rangle$. To verify (iii), suppose that b_i is in $A_i + A_i$ for $i < 4$ and that $\sum_{i<4} b_i = 0$. Certainly

(3) $\quad b_i \in \bar{A}_i + \bar{A}_i \quad \text{for } 1 \leq i \leq 3$,

by definition of the sets A_i. If we can prove that

(4) $\quad b_0 \in \bar{A}_0 + \bar{A}_0$,

then the conclusion $b_i = 0$ for $i < 4$ will follow from condition (iii) for $\langle \bar{A}_i : i < 4\rangle$. Since $\sum_{i<4} b_i = 0$, we have $b_0 = -b_1 - b_2 - b_3$. Thus, b_0 is in B by (1)–(3). Now

(5) $\quad A_0 + A_0 = (\bar{A}_0 + \bar{A}_0) \cup (n + \bar{A}_0) \cup (-n + \bar{A}_0) \cup \{-2n, 2n\}$

by the definition of A_0, the distributivity of $+$ over $\cup$, and the fact that 0 is in $\bar{A}_0 + \bar{A}_0$. Since b_0 is in the left-hand side of (5), it must be in the right-hand side. Suppose that b_0 is in $n + \bar{A}_0$, say $b_0 = n + m$, where $m \in \bar{A}_0$. Then $n = b_0 - m$. Since $b_0 - m$ is in $B + \bar{A}_0$ and n is bigger than all numbers in this set, we have a contradiction. The assumption that b_0 is in $-n+\bar{A}_0$ leads to a similar contradiction. If $b_0 = -2n$ then $-2n$ is in B. Hence, $2n$ is in B by (2). But n — and therefore $2n$ — is bigger than all numbers in this set, by (2) and the choice of n. Again we have a contradiction. The case $b_0 = 2n$ is completely analogous. In view of (5), the only remaining possibility is (4). □

As is now clear, much of the proof of 2.6 falls away in the case when $\mathfrak{G} = \mathfrak{Z}$. Because $g^2 = h^2$ iff $g = \pm h$, the sets X and H (which in this case are ω and $\{0\}$ respectively) are not needed, obviating the need for (1)–(3) and their proof in 2.6. Because $\mathfrak{Z}$ is commutative, the permutations π are not needed and the definition of B simplifies considerably. Consequently, the arguments involved in steps (7)–(12) of 2.6 reduce to a few obvious observations. This reveals (13) to be the key step in the demonstration, and in the commutative case its proof, too, is quite simple.

The conclusions of Lemma 2.6 can be strengthened in two ways. Neither of these strengthenings is needed in the sequel, so we formulate them as remarks. The first strengthening extends 2.6 to the case when the group may be finite. In this case we must also compute upper bounds on the cardinalities of the sets A_i in order to push through the induction.

REVISED VERSION OF LEMMA 2.6. *Let n be a positive integer and let $\mathfrak{G}$ be a group of cardinality at least $2^{300}n^{195}$. Then there is a finite, cyclic subgroup H of $\mathfrak{G}$ such that, for each sequence $m = \langle m_i : i < 4\rangle$ of natural numbers less than n, there is a sequence $A = \langle A_i : i < 4\rangle$ of subsets of G satisfying the following conditions*:

(i) $H \subseteq A_i$ *and* $m_i \leq |A_i| \leq (2m_i + 1) \cdot |H|$ *for* $i < 4$;
(ii) $A_i = A_i^{-1} = A_i \circ H$ *for* $i < 4$;

(iii) *Whenever* $b_i \in A_i^2$ *for* $i < 4$, *and* π *is a permutation of* 4 *such that*

$$b_{\pi(0)} \circ b_{\pi(1)} \circ b_{\pi(2)} \circ b_{\pi(3)} \in H,$$

then $b_i \in H$ *for each* $i < 4$.

PROOF. The proof proceeds much as the proof of Lemma 2.6. However, we must pay special attention to the cardinalities of the sets involved, and this complicates the argument somewhat. The justification for the specific cardinalities that we use will become clear when we construct the sequence A. Let $\equiv$ be the equivalence relation defined in the proof of 2.6.

(I) If $\equiv$ has at least $2^{20}n^{13}$ many equivalence classes, take X to be a set of representatives for the equivalence classes, and take H to be $\{e\}$.

(II) If $\equiv$ has fewer than $2^{20}n^{13}$ many equivalence classes, then one equivalence class must have at least $2^{280}n^{182}$ elements in it (since $|G| \geq 2^{300}n^{195}$); take X to be some such class, and take H to be the subgroup of $\mathfrak{G}$ generated by $\hat{g}^2$ for some fixed $\hat{g} \in X$.

We now establish modified versions of properties (1) and (3) for X and H. Property (2) and its proof remain unchanged.

(1) In case (I) we have $|X| \geq 2^{20}n^{13}$ and $|H| = 1$, and in case (II) we have $|X| \geq 2^{280}n^{182}$ and $|H| < 2^{20}n^{13}$.

(3) Let B, D be arbitrary subsets of G subject to the following cardinality restrictions: $|B| \leq 2^9 n^6$ and $|D| \leq 2n$ in case (I), and $|B| \leq 2^{129}n^{84}$ and $|D| \leq 2^{21}n^{14}$ in case (II). Then there are at least $2n \cdot |H|$ many elements g in X such that $g \notin B \circ D \circ B$ and $g^2 \notin B \sim H$.

In case (I), property (1) is obvious. For (3), observe that

$$|B \circ D \circ B| \leq |B|^2 \cdot |D| \leq 2^{19}n^{13}.$$

Therefore,

$$|X \sim B \circ D \circ B| \geq 2^{20}n^{13} - 2^{19}n^{13} = 2^{19}n^{13}.$$

Because $g^2 \neq h^2$ for distinct g, h in X, there are fewer than $|B|$ many elements g in $X \sim B \circ D \circ B$ such that $g^2 \in B \sim H$. Thus, $X \sim B \circ D \circ B$ has at least $2^{19}n^{13} - |B|$ many g such that $g^2 \notin B \sim H$. Because $|H| = 1$, we obtain (3) in this case.

In case (II), we have $|X| \geq 2^{280}n^{182}$ by definition. As in the proof of 2.6, each element of H has the form h^2 for some $h \in G$. Thus, if we had $|H| \geq 2^{20}n^{13}$, then $\equiv$ would have at least $2^{20}n^{13}$ many equivalence classes, in contradiction to the case that we are in. This proves (1). To prove (3), observe that

$$|B \circ D \circ B| \leq |B|^2 \cdot |D| \leq 2^{279}n^{182}.$$

Therefore,

$$|X \sim B \circ D \circ B| \geq 2^{280}n^{182} - 2^{279}n^{182} = 2^{279}n^{182}.$$

For each $g \in X$, we have $g^2 \in H$ and therefore $g^2 \notin B{\sim}H$. Because

$$2n \cdot |H| < 2n \cdot 2^{20}n^{13} < 2^{279}n^{182}\,,$$

we again arrive at the desired conclusion.

We now proceed exactly as in the proof of 2.6 to construct $\langle A_i : i < 4\rangle$ (by induction on $p = m_0 + m_1 + m_2 + m_3$) so that properties 2.6(i)–(iii), (4), (5), and (9) hold. We must also verify that we always have

(*) $|A_i| \leq (2m_i + 1) \cdot |H|$ for $i < 4$,

and that at each inductive step there is a g satisfying (7). When $p = 0$ we have $A_i = H$, so that (*) is obvious. To verify the existence of a g satisfying (7) in the inductive step we must consider two cases. In both cases we repeatedly use the induction hypothesis (*), property (1), and the assumption $m_0 < n$.

In case (I) we have

$$|\bar{A}_i| \leq (2m_i + 1) \cdot |H| = 2m_i + 1 < 2n$$

for $1 \leq i \leq 3$. Therefore, using the definition (6) we compute:

$$\begin{aligned}|B| &\leq 3! \cdot |\bar{A}_1|^2 \cdot |\bar{A}_2|^2 \cdot |\bar{A}_2|^2 \\ &< 3! \cdot (2n)^2 \cdot (2n)^2 \cdot (2n)^2 \\ &< 2^3 2^6 n^6 \\ &= 2^9 n^6\,.\end{aligned}$$

Also,

$$|\bar{A}_0| \leq [2(m_0 - 1) + 1] \cdot |H| = 2m_0 - 1 < 2n\,.$$

Invoking (3) with $D = \bar{A}_0$, we see that there must by at least $2n$ many elements g in X satisfying the second and third conditions in (7). Since $|\bar{A}_0| < 2n$, one of these g must also satisfy the first condition in (7).

In case (II) we have

$$|\bar{A}_i| \leq (2m_i + 1) \cdot |H| < 2n \cdot 2^{20}n^{13} = 2^{21}n^{14}$$

for $1 \leq i \leq 3$. Using the definition (6), we again compute:

$$\begin{aligned}|B| &\leq 3! \cdot |\bar{A}_1|^2 \cdot |\bar{A}_2|^2 \cdot |\bar{A}_2|^2 \\ &< 3! \cdot (2^{21}n^{14})^2 \cdot (2^{21}n^{14})^2 \cdot (2^{21}n^{14})^2 \\ &< 2^3 \cdot (2^{21}n^{14})^6 \\ &= 2^{129}n^{84}\,.\end{aligned}$$

Also,

$$|\bar{A}_0| \leq [2(m_0 - 1) + 1] \cdot |H| < 2n \cdot 2^{20}n^{13} = 2^{21}n^{14}\,.$$

Invoking again (3), we obtain a g satisfying (7).

It remains to verify (*). When $1 \leq i \leq 3$, this follows from the induction hypothesis (*) on $\bar{A}_i$ and from the definition of A_i. When $i = 0$, we use (8) and the induction hypothesis (*) on $\bar{A}_0$ to obtain:

$$|A_0| \leq |\bar{A}_0| + 2 \cdot |H| \leq [2(m_0 - 1) + 1] \cdot |H| + 2 \cdot |H| = (2m_0 + 1) \cdot |H| .$$

The rest of the proof of the revised lemma proceeds exactly as the proof of 2.6. □

The second strengthening of 2.6 concerns the way in which successive sequences A can be chosen.

ADDENDUM TO LEMMA 2.6. *Given sequences m and A satisfying* 2.6(i)–(iii), *and given another sequence m' of natural numbers such that $m_i \leq m'_i$ for $i < 4$, we can choose the sequence A' for m' so that $A_i \subseteq A'_i$ for $i < 4$.*

PROOF. Indeed, suppose that A has been constructed for m, and that $m_i \leq m'_i$ for $i < 4$. Let $\langle m^{(k)} : k \leq \ell \rangle$ be any sequence of quadruples $m^{(k)} = \langle m_{k,i} : i < 4 \rangle$ of natural numbers such that $m = m^{(0)}$, $m' = m^{(\ell)}$, and

$$\textstyle\sum_{i<4} m_{k+1,i} = (\sum_{i<4} m_{k,i}) + 1 .$$

Set $A^{(0)} = A$. By repeated application of step (9) in the proof of 2.6 we can construct sequences $A^{(k)} = \langle A_{k,i} : i < 4 \rangle$ satisfying 2.6(i)–(iii) for $m^{(k)}$, and with the additional property that $A_{k,i} \subseteq A_{k+1,i}$ for $i < 4$. Then $A^{(\ell)}$ is the desired sequence A' for m'. □

This concludes our remarks regarding extensions of Lemma 2.6. We will refer to these extensions in the remarks following Lemma 2.7. □

LEMMA 2.7. *Let $\mathfrak{G} = \langle G , \circ , ^{-1} , e \rangle$ be an infinite group. Then for every $n \in \omega$ there is a 4-frame in $\mathfrak{Cm}(\mathfrak{G})$ consisting of finite subsets of G such that $\mathfrak{S}(n)$ is embeddable into the associated frame semigroup.*

PROOF. Let H be the finite, cyclic subgroup of $\mathfrak{G}$ given by Lemma 2.6. Fix a natural number $n \geq 1$, and let $\langle A_i : i < 4 \rangle$ be the sequence satisfying conditions 2.6(i)–(iii) for the quadruple

$$\langle n \cdot |H| , n \cdot |H| , n \cdot |H| , n \cdot |H| \rangle$$

of natural numbers.

For each $i < 4$, we define an auxiliary sequence $\langle a_{ip} : p < n \rangle$ of elements in A_i such that

(1) $\quad a_{ip}^{-1} \circ a_{iq} \notin H \quad$ for all distinct $p, q < n$.

We may choose a_{i0} arbitrarily. (For example, set $a_{i0} = e$. This is in A_i, since $H \subseteq A_i$.) Now suppose that $0 < q < n$ and that a_{ip} has been defined for $p < q$. We take a_{iq} to be any element of

$$A_i {\sim} (\{a_{ip} : p < q\} \circ H) .$$

Such an element exists because $\{a_{ip} : p < q\} \circ H$ has at most $q \cdot |H|$ elements, and $|A_i| > n \cdot |H|$, by 2.6(i).

To verify (1), suppose that $p, q < n$ are distinct. If $p < q$, then $a_{iq} \notin a_{ip} \circ H$, by construction. Hence, $a_{ip}^{-1} \circ a_{iq} \notin H$. If $q < p$, then $a_{ip} \notin a_{iq} \circ H$, by construction. Thus, $a_{ip} \neq a_{iq} \circ h$, i.e., $a_{ip}^{-1} \circ a_{iq} \neq h^{-1}$, for every $h \in H$. Since H is a group, the elements h^{-1} vary over all of H as h varies over H. This proves (1).

For each pair of distinct $i, j < n$ and each $R \subseteq n \times n$, set

$$(2) \qquad R_{ij} = \{a_{ip} \circ a_{jq}^{-1} : (p, q) \in R\}.$$

By 2.6(ii) we have

$$(3) \qquad R_{ij} \subseteq A_i \circ A_j .$$

Now fix $R, S \subseteq n \times n$ and distinct $i, j, k, \ell < 4$. (Thus, $\{i, j, k, \ell\} = 4$.) We shall prove that

$$(4) \qquad R_{ij} \otimes S_{jk} = (R|S)_{ik} .$$

Suppose, first, that b is in $R_{ij} \otimes S_{jk}$. Then b is in $(R_{ij} \circ S_{jk}) \cap (A_i \circ A_k)$, by definition of $\otimes$. Therefore, by (2), there must be a pair (p, q) in R, a pair (r, s) in S, a c in A_i, and a d in A_k such that

$$b = a_{ip} \circ a_{jq}^{-1} \circ a_{jr} \circ a_{ks}^{-1} = c \circ d .$$

Trivially,

$$c^{-1} \circ a_{ip} \circ a_{jq}^{-1} \circ a_{jr} \circ a_{ks}^{-1} \circ d^{-1} = e ,$$

so

$$c^{-1} \circ a_{ip} \circ a_{jq}^{-1} \circ a_{jr} \circ a_{ks}^{-1} \circ d^{-1} \in H .$$

Because, e.g., $A_m^{-1} = A_m$ for $m < 4$, and $H \subseteq A_\ell$, we have

$$c^{-1} \circ a_{ip} \in A_i^2 \quad , \quad a_{jq}^{-1} \circ a_{jr} \in A_j^2 \quad , \quad a_{ks}^{-1} \circ d^{-1} \in A_k^2 \quad , \quad e \circ e \in A_\ell^2 .$$

By 2.6(iii), we conclude that $a_{jq}^{-1} \circ a_{jr}$ is in H. Hence, from (1) we obtain $q = r$, and therefore

$$(p, s) \in R|S \qquad \text{and} \qquad b = a_{ip} \circ a_{ks}^{-1} .$$

It now follows from the definition (2) that b is in $(R|S)_{ik}$. This establishes the inclusion from left to right in (4).

For the reverse inclusion, suppose that b is in $(R|S)_{ik}$, say

$$(p, q) \in R \quad , \quad (q, s) \in S \quad , \quad \text{and} \quad b = a_{ip} \circ a_{ks}^{-1} .$$

Since $A_k = A_k^{-1}$, by 2.6(ii), the element b is in $A_i \circ A_k$. Moreover, it is obvious that

$$b = a_{ip} \circ a_{jq}^{-1} \circ a_{jq} \circ a_{ks}^{-1} ,$$

so b is in $R_{ij} \circ S_{jk}$, by (2). Thus, b is in $(R_{ij} \circ S_{jk}) \cap (A_i \circ A_k)$, i.e., it is in $R_{ij} \otimes S_{jk}$. This proves (4).

(5) $\quad R_{ij} \subseteq S_{ij}$ iff $R \subseteq S$.

The implication from right to left in (5) is trivial. For the reverse implication, assume that $R_{ij} \subseteq S_{ij}$, and let (p, q) be in R. Then $a_{ip} \circ a_{jq}^{-1}$ is in R_{ij}, by (2), and hence also in S_{ij}. Thus, there is an (r, s) in S such that

$$a_{ip} \circ a_{jq}^{-1} = a_{ir} \circ a_{js}^{-1} .$$

Trivially,

$$a_{ir}^{-1} \circ a_{ip} \circ a_{jq}^{-1} \circ a_{js} = e ,$$

and therefore,

$$a_{ir}^{-1} \circ a_{ip} \circ a_{jq}^{-1} \circ a_{js} \circ e \circ e \circ e \circ e \in H .$$

Now

$$a_{ir}^{-1} \circ a_{ip} \in A_i^2 \quad , \quad a_{jq}^{-1} \circ a_{js} \in A_j^2 \quad , \quad e \circ e \in A_k^2 \quad , \quad e \circ e \in A_\ell^2 .$$

As before, we conclude from 2.6(iii) that $a_{ir}^{-1} \circ a_{ip}$ and $a_{jq}^{-1} \circ a_{js}$ are in H. Hence, by (1), $p = r$ and $q = s$. Therefore, (p, q) is in S, which proves (5).

Let C be the identity relation on n, i.e., $C = \{(p, p) : p < n\}$. We shall show that

(6) $\quad \langle A_i : i < 4 \rangle \quad , \quad \langle C_{ij} : i, j < 4 \, , \, i \neq j \rangle$ is a 4-frame in $\mathfrak{Cm}(\mathfrak{G})$.

Condition 2.2(i) follows from 2.6(ii). To verify 2.2(ii), let

$$d \in (A_i \circ A_j \circ A_j \circ A_\ell) \cap (A_i \circ A_k \circ A_k \circ A_\ell) .$$

Thus, for each $m < 4$, we can find b_m, b'_m in A_m such that

$$d = b_i \circ b_j \circ b'_j \circ b_\ell = b'_i \circ b_k \circ b'_k \circ b'_\ell .$$

Then

$$b_i'^{-1} \circ b_i \circ b_j \circ b'_j \circ b_\ell \circ b_\ell'^{-1} \circ b_k'^{-1} \circ b_k^{-1}$$

is e, and hence is in H. Invoking 2.6(iii) one more time, we see that $b_j \circ b'_j$ is in H. Because

$$A_i \circ H \circ A_\ell = A_i \circ A_\ell ,$$

by 2.6(ii), we conclude that that

$$b_i \circ b_j \circ b'_j \circ b_\ell \in A_i \circ A_\ell ,$$

i.e., d is in $A_i \circ A_\ell$, as was to be shown.

To verify 2.2(iii), we simply use (4) and the fact that C is the identity relation:

$$(C_{ij} \circ C_{jk}) \cap (A_i \circ A_k) = C_{ij} \otimes C_{jk} = (C|C)_{ik} = C_{ik} .$$

This proves (6).

Next, we shall embed $\mathfrak{S}(n)$ into $\langle L_{01} \, , \odot \rangle$. Let f be the function with domain $Sb\,(n \times n)$ determined by

(7) $\quad f(R) = R_{01} \quad \text{for} \quad R \subseteq n \times n\,.$

Certainly, R_{01} is in L_{01}, by 2.3(i) and (3). For $R, S \subseteq n \times n$ we have

$$\begin{aligned}
f(R) \odot f(S) &= R_{01} \odot S_{01} && \text{by (7),}\\
&= (R_{01} \otimes C_{12}) \otimes (C_{20} \otimes S_{01}) && \text{by 2.3(iii),}\\
&= (R|C)_{02} \otimes (C|S)_{21} && \text{by (4),}\\
&= R_{02} \otimes S_{21} && C \text{ is the identity relation,}\\
&= (R|S)_{01} && \text{by (4),}\\
&= f(R|S) && \text{by (7).}
\end{aligned}$$

Therefore, f is a homomorphism of $\mathfrak{S}(n)$ into $\langle L_{01}\,, \odot\rangle$. By (5), it must be one-one. This completes the proof of the lemma. $\square$

REMARKS. Just as the proof of Lemma 2.6, the proof of Lemma 2.7 simplifies — though not nearly as much — when we assume that $\mathfrak{G} = \mathfrak{Z}$. In this case we may take $\langle a_{ip} : p < n\rangle$ to be any one-one sequence of elements in A_i. Step (1) is obvious since $H = \{0\}$. The proofs of (4)–(6) simplify notationally (though not conceptually) and the proof that f is an embedding remains unchanged.

In general, the function f from the proof of 2.7 does not map $\mathfrak{S}(n)$ *onto* $\langle L_{01}\,, \odot\rangle$. However, f will be an onto mapping if, for $i = 0, 1$, the sequence $\langle a_{ip} : p < n\rangle$ is actually an enumeration of *all* the elements in A_i. Indeed, assume that this is the case and fix an arbitrary X in L_{01}. Continuing the numeration scheme of the proof of 2.7, set

(8) $\quad R_X = \{(p,q) : a_{0p} \circ a_{1q}^{-1} \in X\}\,.$

To show that f is onto, it suffices to show that

(9) $\quad (R_X)_{01} = X\,.$

To prove the inclusion from left to right, suppose that b is in $(R_X)_{01}$. Then by (2) there is a (p,q) in R_X such that $b = a_{0p} \circ a_{1q}^{-1}$. Because (p,q) is in R_X we have b in X by (8). For the reverse inclusion, suppose that b is in X. Then by 2.3.(i) there is a c in A_0 and a d in A_1 such that $b = c \circ d$. Because the sequence $\langle a_{ip} : p < n\rangle$ enumerates all of A_i, and $A_i^{-1} = A_i$, there are $p, q < n$ such that $c = a_{0p}$ and $d^{-1} = a_{1q}$. Hence, $b = a_{0p} \circ a_{1q}^{-1}$. It follows from (8) that (p,q) is in R_X. Therefore, b is in $(R_X)_{01}$ by (2).

The strengthenings of Lemma 2.6 that were given earlier can be used to obtain corresponding strengthenings of Lemma 2.7. We shall not use these strengthenings in the sequel, but they seem to us to be of independent interest. Moreover, the first of them gives an alternate (but more involved) proof of the main result of this chapter, Corollary 2.9. It extends Lemma 2.7 to the case when the group may be finite.

REVISED VERSION OF LEMMA 2.7. *Let $n \in \omega$ and let $\mathfrak{G}$ be a group of cardinality at least $2^{561}(n+1)^{364}$. Then there is a 4-frame in $\mathfrak{Cm}(\mathfrak{G})$ consisting of finite subsets of G such that $\mathfrak{S}(n)$ is embeddable into the associated frame semigroup.*

PROOF. We begin by imitating the proof of the revised version of 2.6 (see the first remark following the proof of 2.6) and combining it with the proof of 2.7. Define $\equiv$ just as before. If $\equiv$ has at least $2^{20}(n+1)^{13}$ equivalence classes, then take X to be a set of representatives for the equivalence classes and set $H = \{e\}$. Taking for m the quadruple

$$\langle n\,,\,n\,,\,n\,,\,n\rangle = \langle n\cdot|H|\,,\,n\cdot|H|\,,\,n\cdot|H|\,,\,n\cdot|H|\rangle\,,$$

we obtain a sequence $A = \langle A_i : i < 4\rangle$ satisfying conditions 2.6(i)–(iii) for this quadruple. We proceed exactly as in the proof of Lemma 2.7 to construct a 4-frame and to embed $\mathfrak{S}(n)$ into the associated frame semigroup.

Suppose, now, that $\equiv$ has fewer than $2^{20}(n+1)^{13}$ equivalence classes. Then some equivalence class has at least $2^{541}(n+1)^{351}$ elements in it (since $|G| \geq 2^{561}(n+1)^{364}$). Take X to be such an equivalence class and take H to be the subgroup of $\mathfrak{G}$ generated by $\hat{g}^2$ for some fixed $\hat{g} \in X$. Exactly as in case (II) of the proof of the revised version of 2.6, one establishes:

(1) $|X| \geq 2^{541}(n+1)^{351}$ and $|H| < 2^{20}(n+1)^{13}$;

(3) Let B, D be arbitrary subsets of G such that $|B| \leq 2^{249}(n+1)^{162}$ and $|D| \leq 2^{41}(n+1)^{27}$. Then there are at least $2^{540}(n+1)^{351}$ many elements g in X such that $g \notin B\circ D\circ B$ and $g^2 \notin B{\sim}H$.

For each quadruple $m = \langle m_i : i < 4\rangle$ with $m_i < (n+1)\cdot|H|$ for each i, we construct, just as before, a sequence $A = \langle A_i : i < 4\rangle$ satisfying the conditions (i)–(iii) in the revised version of 2.6, as well as conditions (4), (5), and (9) in the proof of 2.6. To see that this is possible we need only check that we can always find a g in X satisfying condition (7). At the induction step we have

$$|\bar{A}_i| \leq (2m_i+1)\cdot|H| < 2(n+1)\cdot|H|\cdot|H| < 2(n+1)[2^{20}(n+1)^{13}]^2 = 2^{41}(n+1)^{27}$$

for $1 \leq i \leq 3$. Using the definition (6), we compute:

$$\begin{aligned}|B| &\leq 3!\cdot|\bar{A}_1|^2\cdot|\bar{A}_2|^2\cdot|\bar{A}_2|^2\\ &< 3!\cdot[2^{41}(n+1)^{27}]^2\cdot[2^{41}(n+1)^{27}]^2\cdot[2^{41}(n+1)^{27}]^2\\ &< 2^3\cdot[2^{41}(n+1)^{27}]^6\\ &= 2^{249}(n+1)^{162}\,.\end{aligned}$$

Also,

$$|\bar{A}_0| \leq [2(m_0-1)+1]\cdot|H| < 2(n+1)\cdot|H|\cdot|H| < 2^{41}(n+1)^{27}\,.$$

In view of (3) above, we obtain a g satisfying (7).

Choosing m to be the quadruple $\langle n\cdot|H|\,,\,n\cdot|H|\,,\,n\cdot|H|\,,\,n\cdot|H|\rangle$, we obtain a sequence A satisfying 2.6(i)–(iii). Again, we proceed as in the proof of Lemma 2.7 to construct a 4-frame and to embed $\mathfrak{S}(n)$ into the associated frame semigroup. □

The second remark concerns our ability in Lemma 2.7 to actually embed $\mathfrak{S}(\omega)$ in the associated frame semigroup.

EXTENSION OF LEMMA 2.7. *Let $\mathfrak{G}$ be an infinite group. Then there is a 4-frame in $\mathfrak{Cm}(\mathfrak{G})$ consisting of countable subsets of G such that $\mathfrak{S}(\omega)$ is embeddable into the associated frame semigroup.*

PROOF. To prove this, we continue with the proof of Lemma 2.7. Using Lemma 2.6 and the corresponding addendum (see the second remark following the proof of 2.6), choose, for each $n \in \omega$, a sequence $\langle A_{ni} : i < 4\rangle$ for the quadruple

$$\langle n \cdot |H| \,,\, n \cdot |H| \,,\, n \cdot |H| \,,\, n \cdot |H|\rangle$$

of natural numbers, with the additional stipulation that $A_{mi} \subseteq A_{ni}$ whenever $m < n$. Set $A_i = \bigcup_{n\in\omega} A_{ni}$. It is clear from the definition of the auxiliary sequence, given after (1) (in the proof of 2.7), that an auxiliary sequence for A_{mi} can be extended to an auxiliary sequence for A_{ni} when $m < n$. In this way we construct an auxiliary sequence $\langle a_{ip} : p \in \omega\rangle$ for A_i. Because we have obtained $\langle a_{ip} : p \in \omega\rangle$ by extensions, it is clear from (2) that, for each $R \subseteq \omega$,

$$R_{ij} = \bigcup_{n\in\omega} [R \cap (n \times n)]_{ij} \,.$$

Now the satisfiability of the equations and equivalences in 2.2(i)–(iii) and in (3)–(5) is preserved under unions of chains of sets. Therefore, taking C to be the identity relation on ω, we see from (6), applied to

$$\langle A_{ni} : i < 4\rangle \qquad \text{and} \qquad \langle [C \cap (n \times n)]_{ij} : i, j < 4\rangle \,,$$

that

$$\langle A_i : i < 4\rangle \qquad , \qquad \langle C_{ij} : i, j < 4 \,,\, i \neq j\rangle$$

is a 4-frame; further, the function defined on $Sb(\omega \times \omega)$ by (7) is an embedding of $\mathfrak{S}(\omega)$ into the associated frame semigroup. □

This completes our remarks regarding possible extensions of Lemma 2.7. □

THEOREM 2.8. *Let K be a class of relation algebras. Suppose there is an algebra in K containing an infinite set of pairwise disjoint elements that form a group under ; and $\smile$. Then $\mathcal{E}q(\mathsf{K})$ is recursively inseparable.*

REMARK. In more detail, the hypothesis of the theorem requires the existence of an algebra $\mathfrak{A}$ in K whose universe includes an infinite subset G that satisfies the following conditions: G is closed under the operations ; and $\smile$ of $\mathfrak{A}$, and the equations

$$g^{\smile} \,;\, g = h^{\smile} \,;\, h \qquad , \qquad g^{\smile} \,;\, g \,;\, h = h \qquad , \qquad g \cdot h = 0$$

hold for all g, h in G. In other words, the element $e = g^{\smile} \,;\, g$ is well defined and independent of the choice of g in G, and $\mathfrak{G} = \langle G \,,\, ; \,,\, {}^{\smile} \,,\, e\rangle$ is a group. Furthermore, the elements of G are pairwise disjoint in $\mathfrak{A}$. □

PROOF. Without loss of generality we may assume that K is a variety. Indeed, if the hypothesis holds for K, then it certainly also holds for $\mathbf{HSP}(\mathsf{K})$. Once we show that $\mathcal{E}q(\mathsf{HSP}(\mathsf{K}))$ is recursively inseparable, it follows that $\mathcal{E}q(\mathsf{K})$ is recursively inseparable, since $\mathcal{E}q(\mathsf{K}) = \mathcal{E}q(\mathsf{HSP}(\mathsf{K}))$.

Let ψ, φ, and τ be the formulas and term (introduced before 2.4) that respectively define a 4-frame, the set L_{01}, and the operation $\odot$. We saw in Corollary 2.5 that $\langle \varphi , \tau , \psi \rangle$ is an interpretation of semigroups into relation algebras. We shall now prove that

(1) $\langle \varphi , \tau , \psi \rangle$ interprets finite semigroups into $\mathsf{Si}(\mathsf{K})$.

The theorem follows at once from (1), Lemma 1.5, and Corollary 2.5.

To establish (1), let $\mathfrak{A}$ in K be as in the remark preceding the proof. Thus, there is a set $G \subseteq A$ such that

(2) $\mathfrak{G} = \langle G , ; , {}^{\smile} , e \rangle$ is an infinite group

(where $e = g^{\smile} ; g$ for some g in G), and

(3) $g \cdot h = 0$ for distinct g, h in G.

Without loss of generality,

(4) we may assume that $\mathfrak{A}$ is simple.

Indeed, since K is a variety of relation algebras and G is infinite, $\mathfrak{A}$ is a subdirect product of a non-empty sequence $\langle \mathfrak{B}_\xi : \xi < \kappa \rangle$ of simple algebras in K. Let p_ξ be the projection mapping of $\mathfrak{A}$ onto $\mathfrak{B}_\xi$. Certainly $e \neq 0$, for otherwise we would have $g = g ; e = 0$ for every g in G. Therefore $p_\xi(e) \neq 0$ for some $\xi < \kappa$. Fix such a ξ. Then p_ξ is one-one on G. To see this, let $g, h \in G$. With the help of (2) and the homomorphism property of p_ξ we get

$$p_\xi(g) ; p_\xi(g)^{\smile} = p_\xi(g ; g^{\smile}) = p_\xi(e) \neq 0 .$$

Hence $p_\xi(g) \neq 0$. Suppose that $p_\xi(g) = p_\xi(h)$. Then

$$p_\xi(g \cdot h) = p_\xi(g) \cdot p_\xi(h) = p_\xi(g) \neq 0 .$$

Therefore $g \cdot h \neq 0$. We conclude from (3) that $g = h$. This shows that p_ξ is one-one, and hence an embedding of $\mathfrak{G}$ into the reduct of $\mathfrak{B}_\xi$ to $\{ ; , {}^{\smile} , e\}$. It follows that $p_\xi[G]$ is an infinite set of pairwise disjoint elements in $\mathfrak{B}_\xi$ that form a group under $;$ and ${}^{\smile}$. This proves (4).

The operations $;$ and ${}^{\smile}$ of $\mathfrak{Cm}(\mathfrak{G})$ are the complex operations of the group $\mathfrak{G}$. Thus, for $X, Y \subseteq G$ we have (by (2))

(5) $X ; Y = \{g ; h : g \in X , h \in Y\}$, $X^{\smile} = \{g^{\smile} : g \in X\}$,

where $;$ and ${}^{\smile}$ are performed in $\mathfrak{A}$.

Unfortunately, the elements of $\mathfrak{Cm}(\mathfrak{G})$ are subsets of A, not elements of A. However, we are only concerned with finite subsets, and these can be faithfully interpreted into A. Let f be the function defined on the collection of finite subsets of G by the stipulation

(6) $\quad f(X)=\sum^{\mathfrak{A}} X \quad$ for every finite $X \subseteq G$.

From (3) we readily see that f is one-one. It also preserves Boolean products, relative products, and converses. Indeed, for finite $X, Y \subseteq G$ we have:

$$\begin{aligned}
f(X \cap Y) &= \textstyle\sum(X \cap Y) && \text{by (6)},\\
&= \textstyle\sum\{g \cdot h : g \in X\,,\, h \in Y\} && \text{by (3)},\\
&= (\textstyle\sum X)\cdot(\textstyle\sum Y) && \text{by the distributivity of } \cdot \\
& && \text{over } + \text{ in } \mathfrak{A},\\
&= f(X)\cdot f(Y) && \text{by (6)};
\end{aligned}$$

$$\begin{aligned}
f(X\,;\,Y) &= \textstyle\sum(X\,;\,Y) && \text{by (6)},\\
&= \textstyle\sum\{g\,;h : g \in X\,,\, h \in Y\} && \text{by (5)},\\
&= (\textstyle\sum X)\,;(\textstyle\sum Y) && \text{by the distributivity of } ; \\
& && \text{over } + \text{ in } \mathfrak{A},\\
&= f(X)\,;f(Y) && \text{by (6)};
\end{aligned}$$

$$\begin{aligned}
f(X^{\smile}) &= \textstyle\sum(X^{\smile}) && \text{by (6)},\\
&= \textstyle\sum\{g^{\smile} : g \in X\} && \text{by (5)},\\
&= (\textstyle\sum X)^{\smile} && \text{by the distributivity of } {}^{\smile} \\
& && \text{over } + \text{ in } \mathfrak{A},\\
&= f(X)^{\smile} && \text{by (6)}.
\end{aligned}$$

Fix $n \in \omega$. By (2) and Lemma 2.7, there are sequences

$$\langle A_i : i < 4\rangle \quad , \quad \langle C_{ij} : i,j < 4\,,\, i \neq j\rangle$$

of finite subsets of G forming a 4-frame in $\mathfrak{Cm}(\mathfrak{G})$ such that $\mathfrak{S}(n)$ is embeddable into the associated frame semigroup, $\mathfrak{L}_{01}(\mathfrak{Cm}(\mathfrak{G}))$. Notice that the elements of $\mathfrak{L}_{01}(\mathfrak{Cm}(\mathfrak{G}))$ are also finite subsets of G: they are all below $A_0 \circ A_1$, and this latter set is finite. Thus, all the sets in the frame and in the associated frame semigroup are in the domain of f. Furthermore, the formulas and term ψ, φ, and τ involve only $;$, $\cdot$, and $^{\smile}$. Therefore, they are preserved by f. It follows that

$$\langle f(A_i) : i < 4\rangle \quad , \quad \langle f(C_{ij}) : i,j < 4\,,\, i \neq j\rangle$$

is a 4-frame in $\mathfrak{A}$, and f embeds $\mathfrak{L}_{01}(\mathfrak{Cm}(\mathfrak{G}))$ into $\mathfrak{L}_{01}(\mathfrak{A})$. Consequently, $\mathfrak{S}(n)$ is also embeddable into $\mathfrak{L}_{01}(\mathfrak{A})$. In view of (4), this completes the verification of (1). □

COROLLARY 2.9. *Let* K *be a class of relation algebras. Suppose that, for every* $n \in \omega$, *there is an algebra in* K *containing a set of more than* n *pairwise disjoint elements that form a group under* $;$ *and* $^{\smile}$. *Then* $\mathcal{E}q(\mathsf{K})$ *is recursively inseparable.*

PROOF. By assumption, there is a sequence $\langle \mathfrak{A}_n : n \in \omega\rangle$ of algebras in K and a sequence $\langle G_n : n \in \omega\rangle$ of sets such that, for every $n \in \omega$,

(1) $G_n \subseteq A_n$ and $|G_n| > n$,

(2) $\mathfrak{G}_n = \langle G_n\,, ;^{\mathfrak{A}_n}\,, {}^{\smile\mathfrak{A}_n}\,, e_n \rangle$ is a group (where $e_n = g^{\smile} ; g$ for some $g \in G_n$),

(3) $g \cdot^{\mathfrak{A}_n} h = 0$ for distinct $g, h \in G_n$.

Let $\mathfrak{A}'_n$ be the expansion of $\mathfrak{A}_n$ obtained by adding G_n as a new unary relation, in symbols $\mathfrak{A}'_n = \langle \mathfrak{A}_n\,, G_n \rangle$. Take F to be any non-principal ultrafilter on ω, and define $\mathfrak{B}$, $\mathfrak{B}'$, and H to be the ultraproducts

$$\mathfrak{B} = \left(\prod_{n\in\omega} \mathfrak{A}_n\right)/F \quad , \quad \mathfrak{B}' = \left(\prod_{n\in\omega} \mathfrak{A}'_n\right)/F \quad , \quad H = \left(\prod_{n\in\omega} G_n\right)/F\,.$$

It is easy to check that $\mathfrak{B}' = \langle \mathfrak{B}\,, H \rangle$ (see, e.g., Chang-Keisler [1973], Theorem 4.1.8). The properties formulated in (1)–(3) are first-order expressible in the theory of the structures $\mathfrak{A}'_n$ (see the remarks preceeding the proof of Theorem 2.8) and hence are preserved under the passage to the ultraproduct $\mathfrak{B}'$ (see, *op. cit.*, Theorem 4.1.9). Thus, H is an infinite set of pairwise disjoint elements of $\mathfrak{B}$ that form a group under the operations ; and $^{\smile}$ of $\mathfrak{B}$.

We can now apply Theorem 2.8 to the class $\mathsf{K} \cup \{\mathfrak{B}\}$ to conclude that $\mathcal{E}q(\mathsf{K} \cup \{\mathfrak{B}\})$ is recursively inseparable. Since $\mathfrak{B}$ is an ultraproduct of algebras in K, an equation that is valid in K must also be valid in $\mathfrak{B}$. In other words, $\mathcal{E}q(\mathsf{K}) = \mathcal{E}q(\mathsf{K} \cup \{\mathfrak{B}\})$. Thus, $\mathcal{E}q(\mathsf{K})$ is recursively inseparable. □

REMARKS. First of all, the reader should notice that Theorem 2.8 is actually just a special case of Corollary 2.9.

Secondly, using the revised version of Lemma 2.7 (see the first remark after the proof of the lemma) we could have proved Corollary 2.9 directly in Theorem 2.8, referring to the revised version of Lemma 2.7 instead of to Lemma 2.7 itself. □

COROLLARY 2.10. *Let* K *be any class of relation algebras satisfying one of the following conditions*:

(i) $\mathfrak{Cm}(\mathfrak{G}) \in \mathsf{K}$ *for some infinite group* $\mathfrak{G}$;
(ii) $\mathfrak{Cf}(\mathfrak{G}) \in \mathsf{K}$ *for some infinite group* $\mathfrak{G}$;
(iii) $\{\mathfrak{Cm}(\mathfrak{G}) : \mathfrak{G} \in \mathsf{G}\} \subseteq \mathsf{K}$ *for some infinite collection* G *of pairwise non-isomorphic finite groups.*

Then $\mathcal{E}q(\mathsf{K})$ *is recursively inseparable and hence undecidable.*

PROOF. In both $\mathfrak{Cm}(\mathfrak{G})$ and $\mathfrak{Cf}(\mathfrak{G})$ the set of singletons, $\{\{g\} : g \in G\}$, is a subset of pairwise disjoint elements that form a group under the operations ; and $^{\smile}$. Thus, by Theorem 2.8 in the case of conditions (i) and (ii), and by Corollary 2.9 in the case of condition (iii), $\mathcal{E}q(\mathsf{K})$ is recursively inseparable. In particular, it is not recursive. □

COROLLARY 2.11. *Each of the following classes of relation algebras has a recursively inseparable (and hence undecidable) equational theory.*

(i) *All relation algebras.*
(ii) *Integral relation algebras.*
(iii) *Abelian relation algebras.*
(iv) *Symmetric relation algebras.*

(v) *The subclasses of representable algebras in* (i)–(iv).
(vi) *The subclasses of finite algebras in* (i)–(iv).
(vii) *The subclasses of finite, representable algebras in* (i)–(iv).
(viii) *The subclasses of algebras that are representable over finite sets in* (i)–(iv).

PROOF. Let $\mathfrak{G}_n$ be the Boolean group of order 2^n. Then $\mathfrak{Cm}(\mathfrak{G}_n)$ is well known to be symmetric and simple — and hence also Abelian and integral — and representable over a finite set (using the Cayley representation of the underlying group). Thus, the set $\{\mathfrak{Cm}(\mathfrak{G}_n) : n \in \omega\}$ is included in each of the classes described in (i)–(viii). The corollary now follows from Corollary 2.10(iii). $\square$

Similar methods easily yield:

COROLLARY 2.12. *Each of the following group relation algebras, or classes of group relation algebras, has a recursively inseparable (and hence undecidable) equational theory.*

(i) *All group relation algebras.*
(ii) *Abelian group relation algebras.*
(iii) *p-group relation algebras, for each prime p.*
(iv) *Abelian p-group relation algebras, for each prime p; in particular, Boolean group relation algebras.*
(v) *The subclasses of finite algebras in* (i)–(iv).
(vi) $\{\mathfrak{Cm}(\mathfrak{Z}_n) : 0 < n < \omega\}$.
(vii) $\{\mathfrak{Cm}(\mathfrak{Z}_p) : p$ *a prime*$\}$.
(viii) $\{\mathfrak{Cm}(\mathfrak{Z}_{p^n}) : 0 < n < \omega\}$ *for each prime p*.
(ix) $\mathfrak{Cm}(\mathfrak{Z})$, $\mathfrak{Cm}(\mathfrak{Q})$, $\mathfrak{Cm}(\mathfrak{R})$, *and* $\mathfrak{Cm}(\mathfrak{C})$.
(x) $\mathfrak{Cf}(\mathfrak{Z})$, $\mathfrak{Cf}(\mathfrak{Q})$, $\mathfrak{Cf}(\mathfrak{R})$, *and* $\mathfrak{Cf}(\mathfrak{C})$.

Theorem 2.1 can be used to establish the equational undecidability of classes of relation algebras in which one can find simple algebras with lots of subidentity elements. Corollary 2.10 attacks the other extreme: it furnishes a tool for demonstrating the equational undecidability of classes containing certain integral relation algebras. The next Corollary is intended to illustrate the potential of Corollary 2.9 as a tool for investigating classes of relation algebras intermediate between these two extremes.

For each integer $n > 0$, let I_n be the class of simple relation algebras in which 1' is the sum of n atoms. Thus, I_1 is the class of integral relation algebras. We can think of I_n as the class of simple relation algebras of "integrality degree n".

COROLLARY 2.13. *For each integer* $n \geq 1$ *the class* I_n *and its subclasses of representable algebras, finite algebras, finite and representable algebras, and algebras representable over finite sets, have recursively inseparable (and hence undecidable) equational theories.*

PROOF. Fix $n \geq 1$, and for each $k \in \omega$ let $\mathfrak{G}_k$ be a Boolean group with 2^k elements. Since $\mathfrak{Cm}(\mathfrak{G}_k)$ is representable over a finite set (using the Cayley representation of $\mathfrak{G}_k$), we may treat it as a set relation algebra over a finite base set U_k. For $j < n$, let $\mathfrak{Cm}(\mathfrak{G}_{k,j})$ be a copy of $\mathfrak{Cm}(\mathfrak{G}_k)$ with a finite base set $U_{k,j}$ such that $U_{k,i}$ and $U_{k,j}$ are disjoint when $i \neq j$. We define $\mathfrak{A}_k$ to be the subalgebra of

$\mathfrak{R}(\bigcup_{j<n} U_{k,j})$ generated by the union of the universes of $\mathfrak{Cm}(\mathfrak{G}_{k,0}), \ldots, \mathfrak{Cm}(\mathfrak{G}_{k,n-1})$. (The algebra $\mathfrak{A}_k$ is what Jónsson [1988] calls the *semi-product* — and what Givant [1994] calls a *simple closure* — of the algebras $\mathfrak{Cm}(\mathfrak{G}_{k,0}), \ldots, \mathfrak{Cm}(\mathfrak{G}_{k,n-1})$. That is to say, $\mathfrak{A}_k$ has the following properties: (1) $\mathfrak{A}_k$ is simple; (2) $\langle G_{k,j} : j < n\rangle$ is a sequence of pairwise disjoint equivalence elements in $\mathfrak{A}_k$ whose sum includes the identity element; (3) the relativization of $\mathfrak{A}_k$ to $G_{k,j}$ is just $\mathfrak{Cm}(\mathfrak{G}_{k,j})$ for each $j < n$; and (4) $\mathfrak{A}_k$ is the smallest relation algebra with these properties — see Figure 2.8, and see Givant [1994], pp. 93–94, for a discussion of simple closures.) Since, for each $j < n$, the algebra $\mathfrak{Cm}(\mathfrak{G}_{k,j})$ is integral and representable over a finite set, we see that $\mathfrak{A}_k$ is in I_n and representable over a finite set. Moreover, the set of atoms of $\mathfrak{Cm}(\mathfrak{G}_{k,0})$ is a subset of A_k of cardinality $> k$ that is a group under the operations ; and $\smile$ of $\mathfrak{A}_k$. Thus, we may apply Corollary 2.9 to conclude that I_n and the related subclasses mentioned in the statement of the corollary have undecidable equational theories. □

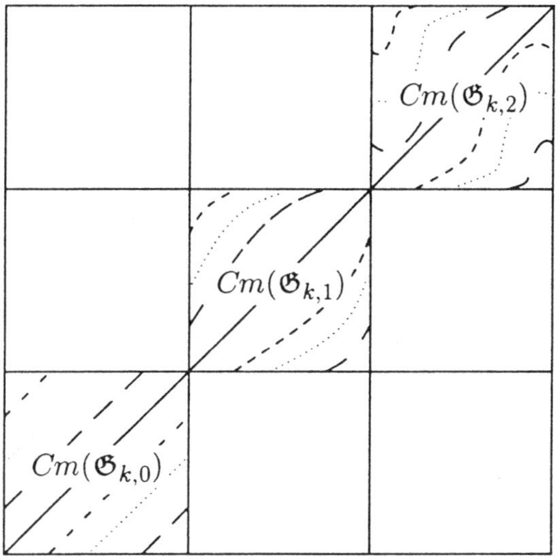

FIGURE 2.8. A picture of $\mathfrak{A}_k$ for the case $n = 3$.

Suppose K is a class of relation algebras satisfying the hypothesis of Corollary 2.9 (e.g., K is one of the classes described in Corollaries 2.11 and 2.12). Because $\mathcal{E}q(\mathsf{K})$ is recursively inseparable, every *set* Θ of equations (and not just every equational theory Θ) satisfying

$$\mathcal{E}q(\mathsf{RA}) \subseteq \Theta \subseteq \mathcal{E}q(\mathsf{K})$$

is non-recursive. The next corollary makes use of this observation. It answers a question posed by Vaughan Pratt (see the historical remarks at the end of the chapter).

DEFINITION 2.14. Let $\mathfrak{A} = \langle A, +, -, ;, ^\smile, 1'\rangle$ be a relation algebra. The *converse-complement operation* of $\mathfrak{A}$ is the unary operation $\veebar$ determined by

$$\veebar x = -(x^\smile) \quad \text{for each } x \in A.$$

Set $\mathfrak{A}_1 = \langle A, +, \cdot, 0, 1, ;, \veebar, 1' \rangle$ and $\mathsf{RA}_1 = \{\mathfrak{B}_1 : \mathfrak{B} \in \mathsf{RA}\}$. $\mathfrak{A}_1$ is called the *positive converse-complement reduct* of $\mathfrak{A}$. □

REMARK. Although $\mathfrak{A}_1$ is not a reduct of $\mathfrak{A}$ in the technical sense, it *is* a reduct of the definitional extension of $\mathfrak{A}$ obtained by adjoining the (definable) operations $\cdot$ and $\veebar$, and the (definable) constants 0 and 1, to the list of fundamental operations of $\mathfrak{A}$. It is natural to use the term *reduct* in this slightly wider sense, i.e., to apply it also to reducts of definitional extensions. Notice that the constant 0' is definable in $\mathfrak{A}_1$: $0' = \veebar 1'$. For this reason we have not included it in the list of fundamental operations and constants of $\mathfrak{A}_1$. □

COROLLARY 2.15. *$\mathcal{E}q(\mathsf{RA}_1)$ is undecidable.*

PROOF. Fix a relation algebra $\mathfrak{A}$ and suppose that b is an element of $\mathfrak{A}$.

(1) If $b \cdot \veebar b = 0$, then $b = b^{\smile}$.

Indeed, if $b \cdot \veebar b = 0$, then $b \cdot -(b^{\smile}) = 0$, so $b \leq b^{\smile}$. Applying $^{\smile}$ to both sides, we get $b^{\smile} \leq b$ (by the relation algebraic laws: $x^{\smile\smile} = x$, and if $x \leq y$, then $x^{\smile} \leq y^{\smile}$). This proves (1).

Every term σ of the language of RA_1 can also be thought of as a term in the language of relation algebras augmented by the operation symbol $\veebar$ (which is defined as in 2.13(i)). Since $\langle \mathfrak{A}, \veebar \rangle$ is a definitional extension of $\mathfrak{A}$, we permit ourselves to write $\sigma^{\mathfrak{A}}[a]$ (or just $\sigma[a]$) for $\sigma^{\langle \mathfrak{A}, \veebar \rangle}[a]$. Notice that $\sigma^{\mathfrak{A}_1}[a] = \sigma^{\mathfrak{A}}[a]$. Hence, in evaluating terms of RA_1, we can always work inside of $\mathfrak{A}$.

In what follows we shall suppose that σ and τ are terms of $\mathcal{L}_r$ with variables among $v_0, \ldots, v_{n-1}$ and that $a \in A$. Define, by induction on terms, a function f mapping terms of $\mathcal{L}_r$ to terms of the language of RA_1:

$$f(v_i) = v_i \quad , \quad f(-\sigma) = \veebar f(\sigma) \quad , \quad f(\sigma + \tau) = f(\sigma) + f(\tau) \, ,$$
$$f(1') = 1' \quad , \quad f(\sigma^{\smile}) = f(\sigma) \quad , \quad f(\sigma \, ; \tau) = f(\sigma) \, ; f(\tau) \quad .$$

Intuitively, $f(\sigma)$ is obtained from σ by eliminating all occurrences of $^{\smile}$ and by replacing all occurrences of $-$ with $\veebar$.

(2) If $\big(f(\delta) \cdot \veebar f(\delta)\big)[a] = 0$ for every subterm δ of σ, then $f(\sigma)[a] = \sigma[a]$.

To establish (2) we show by induction on terms that, for every subterm δ of σ,

(3) $f(\delta)[a] = \delta[a]$.

For example, suppose that $-\delta$ is a subterm of σ and that (3) holds for δ. Then

$(-\delta)[a]$	$= -(\delta[a])$	by the definition of term evaluation,
	$= -(f(\delta)[a])$	by the induction hypothesis (3),
	$= -\big((f(\delta)[a])^{\smile}\big)$	by (1) and the hypothesis of (2),
	$= \veebar \big(f(\delta)[a]\big)$	by the definition of $\veebar$,
	$= \big(\veebar f(\delta)\big)[a]$	by the definition of term evaluation,
	$= f(-\delta)[a]$	by the definition of f.

We leave the remaining cases in the proof of (3) to the reader. By taking σ for δ in (3) we at once get (2).

We now define a second function g mapping terms of $\mathcal{L}_r$ to terms of the language of RA_1:

$$g(\sigma) = \sum\{1 \,;\, [f(\delta) \cdot \breve{\,} f(\delta)] \,;\, 1 \ :\ \delta \text{ is a subterm of } \sigma\}\,.$$

(4) If $\mathfrak{A}$ is simple, then the following three conditions are equivalent: $g(\sigma)[a] = 1$, $g(\sigma)[a] \neq 0$, $f(\delta) \cdot \breve{\,} f(\delta)[a] \neq 0$ for some subterm δ of σ.

Indeed, using the definition of g, the simplicity of $\mathfrak{A}$, and some simple relation algebraic identities, we see that

$$\begin{aligned}
g(\sigma)[a] \neq 0 \quad &\text{iff} \quad \big(1 \,;\, [f(\delta) \cdot \breve{\,} f(\delta)] \,;\, 1\big)[a] \neq 0 &&\text{for some subterm } \delta \text{ of } \sigma,\\
&\text{iff} \quad \big(f(\delta) \cdot \breve{\,} f(\delta)\big)[a] \neq 0 &&\text{for some subterm } \delta \text{ of } \sigma,\\
&\text{iff} \quad \big(1 \,;\, [f(\delta) \cdot \breve{\,} f(\delta)] \,;\, 1\big)[a] = 1 &&\text{for some subterm } \delta \text{ of } \sigma,\\
&\text{iff} \quad g(\sigma)[a] = 1\,.
\end{aligned}$$

With the help of (4) and (2), we can now prove that

(5) The conditional equation

$$(\sigma = \tau) \rightarrow \big(g(\sigma) + g(\tau) + f(\sigma) = g(\sigma) + g(\tau) + f(\tau)\big)$$

is valid in all relation algebras.

By semisimplicity, it suffices to show that the conditional equation in (5) is valid in all simple relation algebras. Assume, therefore, that $\mathfrak{A}$ is simple, and suppose that $\sigma[a] = \tau[a]$ for a given $a \in {}^nA$. We consider two cases. First, suppose that

$$\big(f(\delta) \cdot \breve{\,} f(\delta)\big)[a] \neq 0$$

for some subterm δ of either σ or τ. Then either $g(\sigma)[a] = 1$ or $g(\tau)[a] = 1$, by (4). Hence,

$$\big(g(\sigma) + g(\tau) + f(\sigma)\big)[a] = 1 = \big(g(\sigma) + g(\tau) + f(\tau)\big)[a]\,.$$

Next, suppose that

$$\big(f(\delta) \cdot \breve{\,} f(\delta)\big)[a] = 0$$

for all subterms δ of σ and τ. Then

$$g(\sigma)[a] = g(\tau)[a] = 0\,,$$

by (4). Hence,

$$\begin{aligned}
\big(g(\sigma) + g(\tau) + f(\sigma)\big)[a] &= f(\sigma)[a] \\
&= \sigma[a] &&\text{by (2)},\\
&= \tau[a] &&\text{by assumption},\\
&= f(\tau)[a] &&\text{by (2)},\\
&= \big(g(\sigma) + g(\tau) + f(\tau)\big)[a] &&.
\end{aligned}$$

This proves (5).

(6) The equivalence

$$(\sigma = \tau) \leftrightarrow \big(g(\sigma) + g(\tau) + f(\sigma) = g(\sigma) + g(\tau) + f(\tau)\big)$$

is valid in all symmetric relation algebras.

Because relation algebras are semisimple and because symmetric relation algebras form an equational class, every symmetric relation algebra is a subdirect product of simple symmetric relation algebras. Thus, it suffices to show that the equivalence in (6) holds in all simple symmetric relation algebras. Assume, therefore, that $\mathfrak{A}$ is simple and symmetric, and that $a \in {}^nA$. For every element b of $\mathfrak{A}$ we have

$$b \cdot \veebar b = b \cdot -b = 0\,,$$

by symmetry. Thus, for every term δ,

$$\big(f(\delta) \cdot \veebar f(\delta)\big)[a] = 0\,.$$

Hence,

$$g(\sigma)[a] = g(\tau)[a] = 0\,,$$

by (4). It follows from this observation and (2) that

$$\begin{aligned}\big(g(\sigma) + g(\tau) + f(\sigma)\big)[a] = \big(g(\sigma) + g(\tau) + f(\tau)\big)[a] \quad &\text{iff} \quad f(\sigma)[a] = f(\tau)[a] \\ &\text{iff} \quad \sigma[a] = \tau[a]\,.\end{aligned}$$

This proves (6).

Let Θ be the set of equations $\sigma = \tau$ such that the equation

(7) $\quad g(\sigma) + g(\tau) + f(\sigma) = g(\sigma) + g(\tau) + f(\tau)$

is in $\mathcal{E}q(\mathsf{RA})$ and let Ψ be the equational theory of symmetric relation algebras.

(8) $\quad \mathcal{E}q(\mathsf{RA}) \subseteq \Theta \subseteq \Psi\,.$

Indeed, suppose that $\sigma = \tau$ holds in all relation algebras. Then so does (7), by (5), so $\sigma = \tau$ is in Θ, by definition. Now suppose that $\sigma = \tau$ is in Θ. Then (7) is in $\mathcal{E}q(\mathsf{RA})$, by definition of Θ. In particular, (7) holds in all symmetric relation algebras. But then $\sigma = \tau$ holds in all symmetric relation algebras, by (6). We conclude that $\sigma = \tau$ is in Ψ. This proves (8).

It follows at once from (8) and 2.11(iv) that Θ is not recursive. Now any decision method for $\mathcal{E}q(\mathsf{RA}_1)$ would lead to a decision method for Θ, since

$$\begin{aligned}\sigma = \tau \text{ is in } \Theta \quad &\text{iff} \quad (7) \text{ is in } \mathcal{E}q(\mathsf{RA}) \\ &\text{iff} \quad (7) \text{ is in } \mathcal{E}q(\mathsf{RA}_1)\,.\end{aligned}$$

(Recall here the remarks following the proof of (1).) Thus, $\mathcal{E}q(\mathsf{RA}_1)$ is undecidable. □

We now wish to give an alternate and rather surprising formulation of Corollary 2.9. We first establish a lemma.

LEMMA 2.16. *Let G be a subset of the universe of some relation algebra, and suppose that G satisfies the following conditions*:

(i) *G is closed under $;$ and $\smile$;*
(ii) *the elements of G are non-zero and pairwise disjoint;*
(iii) *G contains an equivalence element e.*

Then the elements of G form a group under $;$ and $\smile$ with e as the identity element.

PROOF. Let e be an equivalence element of G as guaranteed by condition (iii). We begin by showing that

(1) $(e\,;g)\cdot g \neq 0$ and $(g\,;e)\cdot g\neq 0$ for every element g in G.

To prove (1) we use the following laws from the theory of relation algebras (see, e.g., Chin-Tarski [1951], Theorems 2.4, 2.7, and 1.24):

(2) $0\,;x = x\,;0 = 0\,,$

(3) $(x\cdot y)\,;z \leq (x\,;z)\cdot(y\,;z)$ and $z\,;(x\cdot y)\leq (z\,;x)\cdot(z\,;y)\,,$

(4) $(x\,;y)\cdot z\leq x\,;[(x^{\smile}\,;z)\cdot y]$ and $(x\,;y)\cdot z\leq [(z\,;y^{\smile})\cdot x]\,;y\,.$

Let g be an arbitrary element of G. Since e is an equivalence element (see preliminaries), we have

$$[(e^{\smile}\,;e)\cdot 1']\,;g = (e\cdot 1')\,;g\leq e\,;g\,,$$

by (3). Also,

$$[(e^{\smile}\,;e)\cdot 1']\,;g\leq 1'\,;g = g\,,$$

by (3). Thus,

(5) $[(e^{\smile}\,;e)\cdot 1']\,;g\leq (e\,;g)\cdot g\,.$

Taking $x = z = e$ and $y = 1'$ in the first inequality of (4), and using the fact that $1'$ is the identity element, we get

$$e\leq e\,;[(e^{\smile}\,;e)\cdot 1']\,.$$

The reverse inequality follows from the fact that e is an equivalence element. Thus,

(6) $e = e\,;[(e^{\smile}\,;e)\cdot 1']\,.$

Now $e\,;g$ is in G, by condition (i), and hence different from 0 by (ii). Therefore, from (6) we obtain $e\,;[(e^{\smile}\,;e)\cdot 1']\,;g\neq 0$. But then by (2) we must have

$$[(e\,;e^{\smile})\cdot 1']\,;g\neq 0\,.$$

Combining this with (5), we arrive at the first inequality in (1). The second is established in a completely analogous fashion.

Now G consists of pairwise disjoint elements, by condition (ii), and both $e\,;g$ and $g\,;e$ are in G, by (i). Hence, from (1) we obtain

$$e\,;g = g\,;e = g\,,$$

i.e., e is an identity element for G under $;$. Applying the cycle laws to (1) (see the preliminaries), we see that

$$(g^{\smile} ; g) \cdot e \neq 0 \quad \text{and} \quad (g ; g^{\smile}) \cdot e \neq 0 .$$

Because the elements of G are pairwise disjoint and because $g^{\smile} ; g$ and $g ; g^{\smile}$ are in G, by (i), we arrive at

$$g^{\smile} ; g = g ; g^{\smile} = e .$$

In other words, $g^{\smile}$ is an inverse of g under $;$. Since $;$ is at any rate an associative operation, this completes the proof of the lemma. □

It is clear from Lemma 2.16 that Corollary 2.9 can be formulated equivalently as follows.

COROLLARY 2.17. *Let* K *be a class of relation algebras. Suppose that, for every* $n \in \omega$, *there is an algebra in* K *containing a set of more than* n *pairwise disjoint elements that is closed under* $;$ *and* $^{\smile}$ *and that contains an equivalence element. Then* $\mathcal{E}q(\mathsf{K})$ *is recursively inseparable.*

REMARK. We have seen that, in principle, Theorem 2.1 and Corollary 2.9 are intended to handle opposite situations: Theorem 2.1 can be used when the class K has algebras with many subidentity elements, while Corollary 2.9 is applicable when the algebras in K are integral, i.e., 1' is the only non-zero subidentity element. It is therefore somewhat surprising that Corollary 2.9 actually implies Theorem 2.1.

To show this, let K be a class of relation algebras satisfying the hypothesis of Theorem 2.1. We shall show that it must also satisfy the hypothesis of Corollary 2.9. Fix $n \in \omega$, and let $\mathfrak{A}$ be a simple algebra in K with at least n subidentity elements. Let $\mathfrak{G}$ be any group of cardinality n. Without loss of generality, we may assume that the universe of $\mathfrak{G}$ is n. Take $\mathfrak{G}^*$ to be the Cayley representation of $\mathfrak{G}$ as a group of permutations of n. As we saw in the preliminaries, $\mathfrak{G}^*$ is a subalgebra of the Peircean reduct of $\mathfrak{R}(n)$ and the elements of $\mathfrak{G}^*$ (as binary relations) are pairwise disjoint. Consider the function g from the proof of 2.1. By the final part of the proof of 2.1 and by the subsequent remark, g embeds $\langle Sb\,(n \times n) \,, \mid \,, {}^{-1} \rangle$ into $\langle A \,, ; \,, {}^{\smile} \rangle$. Let R and S be two distinct elements of G^*. Then

$$g(R) = \sum\{x_i \,; 1 \,; x_j : (i,j) \in R\} \quad , \quad g(S) = \sum\{x_i \,; 1 \,; x_j : (i,j) \in S\} ,$$

by 2.1(3). Since R and S are disjoint, we see that their images $g(R)$ and $g(S)$ are also disjoint, by 2.1(2). In other words, the mapping g restricted to G^* is an embedding of $\langle G^* \,, \mid \,, {}^{-1} \rangle$ into $\langle A \,, ; \,, {}^{\smile} \rangle$ that takes distinct elements to pairwise disjoint elements. Put another way, the elements of $\{g(R) : R \in G^*\}$ are pairwise disjoint and form a group under $;$ and $^{\smile}$ of cardinality n. Thus, K satisfies the hypothesis of 2.9. By 2.9, $\mathcal{E}q(\mathsf{K})$ is recursively inseparable. (We are indebted to Roger Maddux for suggesting this remark to us and for pointing out that it follows from the proof of Theorem 2.1.) □

Corollary 2.9 seems to be the strongest known result concerning classes of relation algebras with undecidable equational theories. It implies virtually all the concrete equational undecidability results for relation algebras that are known from the literature. In practice, however, Theorem 2.1 is still quite useful since it often applies and its hypothesis is easy to check. We give an example of such an application.

In Givant [1994], Example 7.9, a construction due to Andréka and Németi is given of two commuting equivalence relations S and T on an infinite set U such that together S and T generate an infinite subalgebra $\mathfrak{C}$ of $\mathfrak{R}(U)$. (The construction of these relations given in *op. cit.* is rather indirect. A simple, elegant, and explicit construction can be found in Maddux [1995].) The proof that $\mathfrak{C}$ is infinite actually shows that S and T generate infinitely many subidentity elements. From this and Theorem 2.1 we immediately obtain the following.

COROLLARY 2.18. *If* K *is any class of relation algebras that contains* $\mathfrak{C}$, *then* $\mathcal{E}q(\mathsf{K})$ *is recursively inseparable. In particular, each of the following classes has an undecidable equational theory.*

(i) *The class of relation algebras generated by a set of equivalence elements.*
(ii) *The class of relation algebras generated by two equivalence elements.*
(iii) *The class of relation algebras generated by two commuting equivalence elements.*
(iv) *The subclasses of* (i)–(iii) *consisting of algebras that are representable.*
(v) *The subclasses of* (i)–(iv) *consisting of algebras whose specified generators are reflexive equivalence elements.*

In connection with the preceding corollary, it is of interest to note that the class of relation algebras generated by a single equivalence element has a decidable equational theory. In fact, Jónsson [1988] proved that there are only finitely many isomorphism types of relation algebras generated by a single equivalence element, and they are all finite and explicitly constructible.

HISTORICAL REMARKS. The equational undecidability of the class of set relation algebras (or, equivalently, of the class of representable relation algebras) was announced in Tarski [1941], p. 88. The equational undecidability of the class of all relation algebras was established by Tarski around 1942, but not announced until Chin [1948], pp. 2–3 (see also Chin-Tarski [1951], pp. 341–342, and Tarski [1953]). A proof of these results was published in Tarski-Givant [1987], Theorem 8.5(xii), pp. 251–256. This latter theorem and Theorem 8.5(viii) of *ibid.* actually establish more:

(1) Any variety of relation algebras containing the full relation algebra on some infinite set, or, equivalently, containing all relation algebras over infinite sets, must have an undecidable equational theory. In fact, the set of all equations in a single variable x that hold in the variety is undecidable.

In a seminar on relation algebras in 1970, Tarski posed the problem whether the equational theory of group relation algebras is undecidable. The negative solution that Givant obtained at that time was presented in the seminar, but never published. The proof, which we now sketch, proceeded by interpreting the word problem for groups into the equational theory of group relation algebras. The set of permutational elements of a relation algebra $\mathfrak{A}$ forms a group $\mathfrak{Gr}(\mathfrak{A})$ under $;$, $\smile$, and 1'. Moreover, if $\mathfrak{A} = \mathfrak{Cm}(\mathfrak{G})$, then $\mathfrak{G} \cong \mathfrak{Gr}(\mathfrak{A})$. Thus, fixing a class K of relation algebras that contains all group relation algebras, we see that $\{\mathfrak{Gr}(\mathfrak{A}) : \mathfrak{A} \in \mathsf{K}\}$ must contain all groups, up to isomorphisms. With every open formula $\varphi(v_0, \ldots, v_{n-1})$ in the language of group theory we associate the following open formula $\hat{\varphi}$:

$$[\bigwedge_{i<n}(v_i^{\smile} \,;\, v_i = 1\text{'} \wedge v_i \,;\, v_i^{\smile} = 1\text{'})] \to \varphi' ,$$

where φ' is obtained from φ by replacing the symbols $\circ$, $^{-1}$, and e for the group operations everywhere with ; , $^{\smile}$, and 1'. It is not hard to show that φ holds in all groups iff $\hat{\varphi}$ holds in K. Since the set of conditional equations true of all groups is not recursive, by Novikov [1955], the set of conditional equations true of K is not recursive. By open formula reducibility, the same holds for the set of equations true of K.

Recently, Aanderaa has shown that the set of conditional equations true of all groups is recursively inseparable from the set of conditional equations falsifiable in the class of finite groups. Using the interpretation just described, it is easy to conclude that the equational theory of the class of full group relation algebras on finite groups is recursively inseparable (in our sense). Notice that this conclusion follows at once from Corollary 2.10(i).

In 1974 Maddux showed that the word problem for semigroups could be interpreted into the equational theory of semi-associative relation algebras. In this way, he proved that the set of all equations true in a class of semi-associative relation algebras that includes RRA is undecidable; see Maddux [1978], Chapter 12. By modifying somewhat his approach, he was able to show in 1986 that the set of all equations in a single variable x that are true in a class of semi-associative relation algebras containing $\mathfrak{R}(\omega)$ is not decidable; see Maddux [1994].

The interpretation in Maddux [1994] is somewhat complicated in order to obtain the undecidability of the set of equations in one variable. Putting aside this aim, we can use a simplified version of the proof of Theorem 2.1 to clarify how semigroups can be interpreted into relation algebras in a straightforward manner and what such an interpretation actually achieves. Let K be any class of relation algebras that contains either the full set relation algebra on some infinite set, or else the full set relation algebras on arbitrarily large finite sets. Then the class K' of relative product reducts of algebras in K is a class of semigroups that contains $\mathfrak{S}(U)$ for some infinite U or for arbitrarily large finite U. It follows from the Gurevich-Lewis Theorem mentioned in the proof of 1.5, and from the remarks following 1.4, that the set Φ of conditional equations true of all semigroups (in a language with one binary operation symbol " ; ") is recursively inseparable from the set Ψ of conditional equations falsifiable in K'. Now Φ is just the set of conditional equations of $\mathcal{L}_r$ that involve only ; and are true of all relation algebras (since the relative product reduct of every relation algebra is a semigroup). Also, Ψ is just the set of conditional equations of $\mathcal{L}_r$ that involve only ; and are falsifiable in K. Therefore, the set of conditional equations true of all relation algebras is recursively inseparable from the set of conditional equations falsifiable in K. By open formula reducibility, the same holds for the analogous sets of equations.

From this we obtain at once the equational undecidability of the class of finite relation algebras, the class of finite representable relation algebras, and the class of relation algebras representable on finite sets (see Corollary 2.11(i),(vi)–(viii)). The equational undecidability of this last class is due to Schönfeld [1979].

Theorem 2.1 itself dates back to around 1988. A talk on the subject was given by the second author in the fall of 1991, during the Semester on Algebraic Logic held in Warsaw, Poland, at the Banach Center. The result was not announced in print until Andréka-Givant-Németi [1994a].

The undecidability of the equational theory of symmetric relation algebras was proved by Jipsen around 1989, using results of Maddux [1981]. His argument actually shows that any class of Abelian relation algebras that contains all simple, symmetric relation algebras in which the equation $x \leq x\,;x$ is valid must have an undecidable equational theory. (See Jipsen [1992], Theorem 3.24.)

The following version of Corollary 2.10 was obtained by the authors in 1991, using the results of Lipshitz [1974].

(2) Let K be any class of Abelian relation algebras that satisfies the condition:

> (α) Either the full group relation algebra over (the group reduct of) a vector space of infinite dimension is in K, or full group relation algebras over vector spaces of arbitrarily large finite dimension are in K.
>
> Then K must have an undecidable equational theory.

(This version of Corollary 2.10, together with Theorem 2.1 and the 1970 result mentioned above, was enough to solve all of Tarski's decision problems for classes of relation algebras.) We communicated this result to Alasdair Urquhart. He responded that, using similar methods (which he first applied in Urquhart [1984]), he had proved a closely related, but still unpublished result: any variety of distributive lattice-ordered semigroups (DL-semigroups) that contains the DL-semigroup of subspaces of some infinite dimensional vector space has a finitely presented DL-semigroup with an unsolvable word problem (see Urquhart [1993]). He pointed out that his methods had allowed him to avoid the assumption of commutativity. We then saw that it was easy to avoid this assumption in our own proof as well. Thus, in (2) K could be an arbitrary class of relation algebras satisfying condition (α). Independently, Herrmann [1993] proved the following theorem: if K is a class of *representable* relation algebras satisfying (α), then the set of conditional equations true of representable relation algebras is recursively inseparable from the set of conditional equations falsifiable in K. Theorem 2.8 and Corollary 2.9, which are much stronger than our earlier result, were obtained by the authors in January 1994 and announced in Andréka-Givant-Németi [1994b].

Andréka-Bredikhin [1994] showed that the positive equational theory of relation algebras — that is to say, the set of equations (not involving complementation) true of all positive reducts,

$$\langle A\,, +\,, \cdot\,, 0\,, 1\,, ;\,, {}^{\smile}\,, 1\text{'}\,, 0\text{'}\rangle\,,$$

of relation algebras $\mathfrak{A}$ — is *decidable*. Motivated by this result and by considerations in computer science, Vaughan Pratt (in a private communication to two of the authors) asked whether the positive converse-complement equational theory of relation algebras — that is to say, $\mathcal{E}q(\mathsf{RA}_1)$ — is decidable. Corollary 2.14 gives a negative answer to Pratt's query.

At the conference Relational Methods in Computer Science II, held in Parati, Brazil in 1995, Ewa Orlowska asked one of the authors if the class of relation algebras that are generated by a set of equivalence elements has a decidable equational theory. Motivated by her question, Peter Jipsen then asked whether the class of relation algebras generated by two equivalence elements had a decidable equational theory. Corollary 2.18 dates from that time. □

CHAPTER III

A LATTICE EMBEDDING THAT PRESERVES DECIDABILITY AND UNDECIDABILITY

The goal of this chapter is to construct an embedding $X \to \mathsf{L}_X$ of the complete lattice $\langle Sb(\omega), \cup, \cap \rangle$ of subsets of ω into an interval of the lattice of varieties of relation algebras such that L_X has a decidable equational theory iff X is decidable, i.e., recursive. In fact, if X is not recursive, then X and $\mathcal{E}q(\mathsf{L}_X)$ will have the same degree of unsolvability. As was mentioned in Chapter 1, the lattice of varieties of relation algebras and the lattice $\mathcal{S}$ of universal classes of simple relation algebras are canonically isomorphic. Thus, it suffices to embed $\langle Sb(\omega), \cup, \cap \rangle$ into $\mathcal{S}$. This is, in fact, what we shall do. The relation algebras that we shall use are Lyndon algebras on projective lines, first introduced in Lyndon [1961] as part of a general class of relation algebras constructed from projective geometries. (Lyndon's construction had its roots in an earlier construction of Jónsson [1959].)

We fix a special element e for the remainder of the chapter. Let U be any set of cardinality at least 3 that does not contain e. We define the *Lyndon algebra on U* to be the algebra

$$\mathfrak{Ln}(U) = \langle Sb(U \cup \{e\}), +, -, ;, \smile, 1' \rangle,$$

where $\langle Sb(U \cup \{e\}), +, - \rangle$ is the Boolean algebra of subsets of $U \cup \{e\}$ under the usual set-theoretic operations, $\smile$ is the identity operation on subsets of $U \cup \{e\}$, 1' is the element $\{e\}$ and $\{e\}$ is the identity element for ; , and ; is the completely distributive binary operation on subsets of $U \cup \{e\}$ that is defined between singletons of $Sb(U)$, i.e., between subdiversity atoms, by the rules

$$\begin{aligned} &\{p\} ; \{p\} = \{p, e\} = \{p\} + 1', \\ &\{p\} ; \{q\} = U \sim \{p, q, e\} = 1 - (\{p\} + \{q\} + 1') \quad \text{for} \quad p \neq q. \end{aligned}$$

Lyndon [1961] showed (and it is not hard to check) that $\mathfrak{Ln}(U)$ is a relation algebra (provided that U is not a two-element set). $\mathfrak{Ln}(U)$ is obviously symmetric and integral (1' is an atom), and its diversity element is U. Therefore, $\mathfrak{Ln}(U)$ and its subalgebras are simple, and relative multiplication is commutative and completely determined by the products of the non-zero subdiversity elements. Some simple calculations produce a complete set of rules (modulo commutativity) for computing the relative product of two such elements. Indeed, given non-empty sets $X, Y \subseteq U$, let

$$Z = X \cdot Y \quad , \quad X' = X - Y \quad , \quad Y' = Y - X.$$

Then X', Y', and Z are pairwise disjoint, and by the distributivity of ; over + we have

$$X ; Y = X' ; Y' + X' ; Z + Y' ; Z + Z ; Z.$$

Thus, to specify the operation of relative multiplication, it suffices to specify the rules for multiplying a non-zero subdiversity element with itself and for multiplying two non-zero, disjoint, subdiversity elements. Let $X, Y \subseteq U$ be non-zero and disjoint. Then

$$X\,;X = \begin{cases} 1 & \text{if} \quad |X| \geq 2, \\ 1\text{'} + X & \text{if} \quad |X| = 1, \end{cases}$$

$$X\,;Y = \begin{cases} 0\text{'} & \text{if} \quad |X|, |Y| \geq 2, \\ -(1\text{'} + X) & \text{if} \quad |X| = 1,\ |Y| \geq 2, \\ -(1\text{'} + X + Y) & \text{if} \quad |X| = |Y| = 1. \end{cases}$$

A simple consequence of these observations is that, for any elements X, Y of $\mathfrak{Ln}(U)$ the product $X\,;Y$ is always a Boolean combination of X, Y, and 1'.

Each one of the relative multiplication rules can actually be expressed by an open formula. To see this, notice first of all that the only equivalence elements in $\mathfrak{Ln}(U)$ are 0, 1', 1, and sets of the form $\{p, e\}$, where $p \in U$. Thus, the following open formula, which we denote by $\sigma(x)$, characterizes singleton subsets of U:

$$[0 < x < 0\text{'}] \wedge [(x + 1\text{'})\,;(x + 1\text{'}) \leq x + 1\text{'}]\,.$$

The rule

$$X\,;Y = -(1\text{'} + X) \quad \text{if} \quad |X| = 1\,,\ \ |Y| \geq 2\,,\ \ X \cdot Y \doteq 0\,, \quad \text{and} \quad X, Y \leq 0\text{'}$$

can be expressed by the open formula

$$[\sigma(x) \wedge (0 < y \leq 0\text{'}) \wedge -\sigma(y) \wedge (x \cdot y = 0)] \rightarrow [x\,;y = -(1\text{'} + x)]\,.$$

The other rules are handled in a similar fashion. Taking the conjunction of the open formulas that express the various rules, we obtain an open formula $\varphi(x, y)$ that completely characterizes relative multiplication in a Lyndon algebra. Let Λ be the set of formulas consisting of the equations axiomatizing the class of relation algebras, the open formula expressing the property of being simple (see the preliminaries), the equation $x^{\smile} = x$ expressing symmetry, the inequality $1\text{'} \neq 1$, and φ.

Define L to be the class of algebras embeddable into Lyndon algebras. Thus, $\mathfrak{A} \in \mathsf{L}$ iff there is a set U of cardinality at least 3 (with $e \notin U$) such that $\mathfrak{A}$ is embeddable into $\mathfrak{Ln}(U)$. We now describe a class of algebras in L. Let (κ, λ) be any pair of cardinals with $\kappa \geq 3$ or else $\lambda \geq 1$. Suppose, for the moment, that $(\kappa, \lambda) \neq (0, 1)$. Take sequences

$$\langle a_\xi : \xi < \kappa \rangle \quad , \quad \langle b_\eta : \eta < 2\lambda \rangle$$

of distinct elements, with no term of one sequence occurring in the other, and set

$$S_\kappa = \{\{a_\xi\} : \xi < \kappa\} \quad \text{and} \quad T_\lambda = \{\{b_{2\eta}, b_{2\eta+1}\} : \eta < \lambda\}\,.$$

Define $\mathfrak{L}_{\kappa\lambda}$ to be the complete subalgebra of the Lyndon algebra on the set

$$\{a_\xi : \xi < \kappa\} \cup \{b_\eta : \eta < 2\lambda\}$$

that is generated by $S_\kappa \cup T_\lambda$. (Notice that this really is a Lyndon algebra: since $\kappa \geq 3$ or $\lambda \geq 1$, and $(\kappa, \lambda) \neq (0, 1)$, we must have $\kappa + 2\lambda \geq 3$.) Thus, the atoms of $\mathfrak{L}_{\kappa\lambda}$ are $\{e\}$ and the elements of $S_\kappa \cup T_\lambda$. A set X is in $L_{\kappa\lambda}$ iff there is a $\Gamma \subseteq \kappa$ and a $\Delta \subseteq \lambda$ such that

$$X = \bigcup_{\xi\in\Gamma}\{a_\xi\} \cup \bigcup_{\eta\in\Delta}\{b_{2\eta}, b_{2\eta+1}\}$$

or

$$X = \bigcup_{\xi\in\Gamma}\{a_\xi\} \cup \bigcup_{\eta\in\Delta}\{b_{2\eta}, b_{2\eta+1}\} \cup \{e\}\,.$$

We define $\mathfrak{L}_{01}$ to be the subalgebra of $\mathfrak{Ln}(\{b_0, b_1, b_2\})$ generated by the diversity element $\{b_0, b_1, b_2\}$. Thus, $\mathfrak{L}_{01}$ is just the subalgebra of constants of $\mathfrak{Ln}(\{b_0, b_1, b_2\})$.

THEOREM 3.1. (i) L *is a universal class. In fact,* Λ *is a set of axioms for* L.

(ii) $\mathbf{SP}(\mathsf{L})$ *is locally finite, i.e., finitely generated members are finite. In fact, for each* $\mathfrak{A} \in \mathsf{L}$, *a subset of* A *is a subuniverse of* $\mathfrak{A}$ *iff it contains* 1' *and is closed under the Boolean operations.*

(iii) *Every atomic algebra* $\mathfrak{A}$ *in* L *is embeddable into* $\mathfrak{L}_{\kappa\lambda}$ *for some pair of cardinals* (κ, λ) *with* $\kappa \geq 3$ *or* $\lambda \geq 1$. *If* $\mathfrak{A}$ *is also complete, then we can choose* λ *so that this embedding is an isomorphism.*

(iv) $\mathfrak{A}$ *is a finite algebra in* L *iff* $\mathfrak{A} \cong \mathfrak{L}_{\kappa\lambda}$ *for some pair* (κ, λ) *of finite cardinals with* $\kappa \geq 3$ *or* $\lambda \geq 1$.

(v) *Let* (κ, λ) *and* (μ, ν) *be pairs of cardinals with* $\kappa \geq 3$ *or* $\lambda \geq 1$, *and* $\mu \geq 3$ *or* $\nu \geq 1$.

(α) $\mathfrak{L}_{\kappa 0}$ *is embeddable into* $\mathfrak{L}_{\mu\nu}$ *iff* $\kappa = \mu$ *and* $\nu = 0$; *such an embedding must actually be an isomorphism.*

(β) *For* $\lambda \geq 1$ *and* μ *finite,* $\mathfrak{L}_{\kappa\lambda}$ *is embeddable into* $\mathfrak{L}_{\mu\nu}$ *iff* $\kappa \leq \mu$ *and* $\lambda \leq \lfloor(\mu - \kappa)/2\rfloor + \nu$.

(γ) *For* $\lambda \geq 1$ *and* μ *infinite,* $\mathfrak{L}_{\kappa\lambda}$ *is embeddable into* $\mathfrak{L}_{\mu\nu}$ *iff* $\kappa \leq \mu$ *and* $\lambda \leq \mu + \nu$.

(δ) *If all cardinals are finite, then* $\mathfrak{L}_{\kappa\lambda} \cong \mathfrak{L}_{\mu\nu}$ *iff* $\kappa = \mu$ *and* $\lambda = \nu$.

PROOF. We begin by showing that

(1) Every atomic model of Λ is embeddable into $\mathfrak{L}_{\kappa\lambda}$ for some pair of cardinals (κ, λ) with $\kappa \geq 3$ or $\lambda \geq 1$. If the model is also complete, then we can choose λ so that this embedding is an isomorphism.

To prove (1), suppose that $\mathfrak{A}$ is an atomic model of Λ. Notice that 1' is an atom, since a simple, symmetric relation algebra is integral (see Chapter 1). Partition the subdiversity atoms of $\mathfrak{A}$ into two sets: the set S of those atoms that satisfy $\sigma(x)$, and the set T of those that do not. Set $\kappa = |S|$ and $\lambda = |T|$. We first treat the case when $\kappa \geq 3$ or $\lambda \geq 1$, and $(\kappa, \lambda) \neq (0, 1)$. Let f be any function with domain $S \cup T \cup \{1'\}$

that maps S bijectively to $\{\{a_\xi\} : \xi < \kappa\}$, T bijectively to $\{\{b_{2\eta}, b_{2\eta+1}\} : \eta < \lambda\}$, and 1' to $\{e\}$. Then f maps the atoms of $\mathfrak{A}$ bijectively to the atoms of $\mathfrak{L}_{\kappa\lambda}$, so it can be extended to a completely additive Boolean embedding of A into $L_{\kappa\lambda}$. Moreover, if $\mathfrak{A}$ is complete, then f will be a Boolean isomorphism. Certainly, f preserves the converse operation, since the latter is the identity function in both algebras. Also, f will preserve the relative product of two atoms, since this is completely determined by φ in both algebras. For example, suppose $x \in S$ and $y \in T$. Then

$$\begin{aligned} f(x\,;y) &= f(-(1\text{'} + x)) && \text{since } (x,y) \text{ satisfies } \varphi, \\ &= -[f(1\text{'}) + f(x)] && \text{since } f \text{ is a Boolean embedding,} \\ &= -[\{e\} + f(x)] && \text{since } f(1\text{'}) = \{e\}, \\ &= f(x)\,;f(y) && \text{since } |f(x)| = 1 \text{ and } |f(y)| = 2, \text{ and } \varphi \\ & && \text{holds in } \mathfrak{L}_{\kappa\lambda}\,. \end{aligned}$$

Because f preserves the relative product of atoms and is completely additive, it must preserve arbitrary relative products. (We are using here the fact that every element of $\mathfrak{A}$ is a sum of atoms, and that the relative product operation is completely distributive.) Thus, f is a relation algebraic embedding, and actually an isomorphism when $\mathfrak{A}$ is complete.

In the case when $(\kappa, \mu) = (0, 1)$, we see that $\mathfrak{A}$ has a single subdiversity atom, and this atom is in T. Thus, $\mathfrak{A}$ has just two atoms, 0' and 1', and hence only four elements: 0, 1, 0', 1' . In other words, $\mathfrak{A}$ is an algebra of constants. Moreover, using φ we see that $0\text{'}\,;0\text{'} = 1$ and that $\mathfrak{A}$ is simple. Now $\mathfrak{L}_{01}$ has exactly the same properties. But up to isomorphisms, there is only one simple algebra of constants in which $0\text{'}\,;0\text{'} = 1$. Thus, $\mathfrak{A} \cong \mathfrak{L}_{01}$.

The same argument shows that $\mathfrak{A} \cong \mathfrak{L}_{01}$ when $(\kappa, \lambda) = (1, 0)$, since in this case we have $0\text{'};0\text{'} = 1\text{'}+0\text{'} = 1$. The remaining two cases, where (κ, λ) is $(0, 0)$ or $(2, 0)$, are impossible. Indeed, if $\kappa = 0$ and $\lambda = 0$, then 1 is an atom, and hence $1\text{'} = 1$. But Λ specifies that $1\text{'} \neq 1$. If $\kappa = 2$ and $\lambda = 0$, say $S = \{x, y\}$ and $T = \varnothing$, then x, y, and 1' are the atoms of $\mathfrak{A}$. Therefore, $x + y + 1\text{'} = 1$. But then, using φ, we get that $x\,;y = -(1\text{'} + x + y) = 0$, although x and y are non-zero. This contradicts the fact that $\mathfrak{A}$, being simple and Abelian, must be integral. This completes the proof of (1).

Let K be the class of atomic models of Λ.

(2) $\mathsf{L} = \mathbf{S}(\mathsf{K})$.

Indeed, L contains the algebras $\mathfrak{L}_{\kappa\lambda}$ for $\kappa \geq 3$ or $\lambda \geq 1$ (by the definition of these algebras), and L is closed under formation of subalgebras. Thus, $\mathsf{K} \subseteq \mathsf{L}$, by (1), and therefore $\mathbf{S}(\mathsf{K}) \subseteq \mathsf{L}$. On the other hand, each Lyndon algebra is atomic and is a model of Λ. Thus, K contains all Lyndon algebras. Since every algebra in L is embeddable into a Lyndon algebra, by definition, we see that $\mathsf{L} \subseteq \mathbf{S}(\mathsf{K})$.

(3) L is a universal class.

Certainly, K is first-order axiomatizable, since the property of being an atomic algebra is first-order expressible. Applying the well-known Łoś–Tarski Theorem (see, e.g., Tarski [1954], Theorem 1.6, and the remarks preceding Theorem 2.1 in Tarski [1955]), we obtain (3) directly from (2).

(4) If $\mathfrak{A}$ is a model of Λ, then a subset X of A is a subuniverse of $\mathfrak{A}$ iff it is closed under the Boolean operations of $\mathfrak{A}$ and contains 1'.

To prove the non-trivial direction of (4), assume that X is closed under the Boolean operations of $\mathfrak{A}$ and contains 1'. Certainly, X is closed under conversion, the identity operation on A. The relative multiplication laws expressed in φ imply that for any $x, y \in A$, the product $x\,;y$ is a Boolean combination of x, y and 1'. To see this, set $x = u + z$ and $y = v + z$, where

$$z = x \cdot y \quad , \quad u = x - y \quad , \quad v = y - x\,.$$

Then

$$x\,;y = u\,;v + u\,;z + v\,;z + z\,;z\,.$$

Because u, v, and z are disjoint, we can apply φ and the laws of relation algebra to determine the four summands

$$u\,;v \quad , \quad u\,;z \quad , \quad v\,;z \quad , \quad z\,;z\,.$$

Each of these is a Boolean combination of its factors and 1', as is obvious from the form of φ and from the relation algebraic laws regarding 1' and 0. But the factors u, v, and z are themselves Boolean combinations of x and y. Because X is closed under the Boolean operations, we conclude that $x\,;y$ is in X. This proves (4).

Since a finitely generated Boolean algebra is always finite, we see from (4) that

(5) Every finitely generated model of Λ is finite, and hence atomic.

We turn to the proof of (i), and we begin by showing that every model of Λ is in L. Let $\mathfrak{A}$ be an arbitrary model of Λ. A subalgebra of $\mathfrak{A}$ is a model of Λ, since Λ consists of open formulas. Thus, every finitely generated subalgebra of $\mathfrak{A}$ is an atomic model of Λ, by (5), and therefore in L, by (1). Now $\mathfrak{A}$ is the directed union of its finitely generated subalgebras, each of which is in L. Because L is universal, it is closed under directed unions. Thus, $\mathfrak{A}$ itself is in L. The reverse implication, that every algebra in L is a model of Λ, follows at once from (2): K is a class of models of Λ, by definition, so certainly every algebra in $\mathbf{S}(\mathsf{K})$ is a model of Λ.

For part (ii), we see from (i) and (5) that L is locally finite. Since L is a universal class, $\mathbf{SP}(\mathsf{L})$ is locally finite, by Theorem 1.8. The second assertion in (ii) follows at once from (i) and (4).

Part (iii) is a direct consequence of (i) and (1); and part (iv) follows easily from (iii).

We now turn to the proof of (v). Let (κ, λ) and (μ, ν) be pairs of cardinals such that $\kappa \geq 3$ or $\lambda \geq 1$, and $\mu \geq 3$ or $\nu \geq 1$. We prove (α) and (β) simultaneously, and leave the similar, but easier proofs of (γ) and (δ) to the reader. First, under the assumption that the cardinals satisfy the given conditions, we construct an embedding f of $\mathfrak{L}_{\kappa\lambda}$ into $\mathfrak{L}_{\mu\nu}$. The construction is trivial in the case of (α). Turning to (β), suppose that

$$1 \leq \kappa \leq \mu < \omega \qquad \text{and} \qquad 1 \leq \lambda \leq \lfloor (\mu - \kappa)/2 \rfloor + \nu\,.$$

Let g_0 map S_κ one-one into S_μ. Such a g_0 exists because $\kappa \leq \mu$. Notice that $|\, S_\mu {\sim} g_0[S_\kappa]\, | = \mu - \kappa$. Consider some fixed pairing of the remaining singletons in S_μ (i.e., of the elements of $S_\mu {\sim} g_0[S_\kappa]$); we remark in passing that there will be one remaining singleton left unpaired just in case $\mu - \kappa$ is odd. Take D to be the set of doubletons obtained by combining (i.e., forming the union of) the paired singletons; for example, if $\{x\}$ and $\{y\}$ are paired singletons, then $\{x, y\}$ is one of the elements of D. Then $|D| = \lfloor(\mu - \kappa)/2\rfloor$, and therefore

$$|T_\nu \cup D| = \lfloor(\mu - \kappa)/2\rfloor + \nu\,.$$

Since $\{b_0, b_1\}$ is one of the pairs of T_λ (recall that $\lambda \geq 1$), we obtain

$$|T_\lambda {\sim} \{\{b_0, b_1\}\}| = \lambda - 1 < \lambda \leq \lfloor(\mu - \kappa)/2\rfloor + \nu = |T_\nu \cup D|\,.$$

Thus, there must be a one-one mapping g_1 of $T_\lambda - \{\{b_0, b_1\}\}$ into $T_\nu \cup D$, and there will be at least one element of $T_\nu \cup D$ that is not in the range of g_1. Let x be the union of the remaining elements of $T_\nu \cup D$ (i.e., the elements of $(T_\nu \cup D){\sim} g_1[T_\lambda]$), together with the remaining singleton of S_μ in the case when $\mu - \kappa$ is odd. Take f to be the function from $S_\kappa \cup T_\lambda \cup \{e\}$ (the set of atoms of $\mathfrak{L}_{\kappa\lambda}$) into $L_{\mu\nu}$ that is determined by

(6) $\quad f = g_0 \cup g_1 \cup \{(\{b_0, b_1\}, x)\} \cup \{(\{e\}, \{e\})\}\,.$

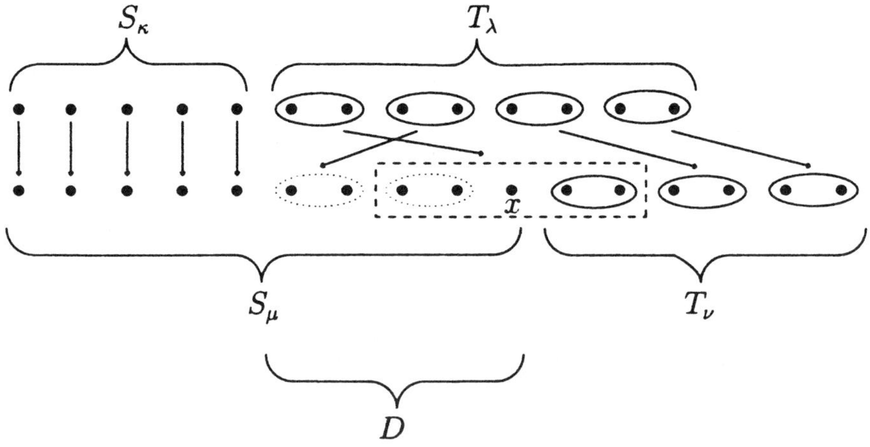

FIGURE 3.1. The function f

It is easy to check that f has the following properties.

(7) $\quad f[S_\kappa \cup T_\lambda]$ is a partition of diversity in $\mathfrak{L}_{\mu\nu}$, i.e., the sum of all of the elements in $f[S_\kappa \cup T_\lambda]$ is 0', and the meet of any two of them is 0.

(8) $\quad f(y) \in S_\mu$ iff $y \in S_\kappa$, and $|f(y)| \geq 2$ iff $y \in T_\lambda$.

In view of (6) and (7), the function f maps the atoms of $\mathfrak{L}_{\kappa\lambda}$ to a partition of the Boolean unit of $\mathfrak{L}_{\mu\nu}$. Therefore, f can be extended in a completely additive way to a Boolean embedding of $\mathfrak{L}_{\kappa\lambda}$ into $\mathfrak{L}_{\mu\nu}$. Certainly, this extension (which we also denote by f) preserves conversion, in this case the identity operation. The arithmetic rules for relative multiplication show that, for two non-zero, subdiversity elements u and v that are equal or disjoint, their relative product $u\,;v$ depends only on $|u|$ and $|v|$, and in fact it depends only on whether these cardinals are 1 or ≥ 2. Therefore, (8) insures us that f will preserve relative multiplication. We conclude that f is a relation algebraic embedding.

To prove the reverse directions of (α) and (β), suppose that f is an embedding of $\mathfrak{L}_{\kappa\lambda}$ into $\mathfrak{L}_{\mu\nu}$. Certainly, f must satisfy properties (7) and (8). (For example, to verify (7), notice that the elements of $S_\kappa \cup T_\lambda$ — arranged in some sequence — form a partition of diversity in $\mathfrak{L}_{\kappa\lambda}$. Since the property of a sequence of elements forming such a partition is expressible by an open formula, it must be preserved by f.) Thus, $\kappa \leq \mu$, by (8). To treat (α), assume that $\lambda = 0$. Then S_κ is a partition of 0' in $\mathfrak{L}_{\kappa\lambda}$, so $f[S_\kappa]$ must be a partition of 0' in $\mathfrak{L}_{\mu\nu}$. Since $f[S_\kappa] \subseteq S_\mu$, by (8), this forces $f[S_\kappa] = S_\mu$ and $T_\nu = \varnothing$; hence, we have $\kappa = \mu$ and $\nu = 0$. To treat (β), assume that $1 \leq \lambda$ and that μ is finite. By (7), the set $f[T_\lambda]$ must form a partition of $\bigcup[(S_\mu {\sim} f[S_\kappa]) \cup T_\nu]$ into λ many sets, and by (8), each of these λ sets has cardinality at least 2. Thus, using also the definition of S_μ and T_ν, we get

$$2\lambda \leq |\bigcup f[T_\lambda]| = |\bigcup[(S_\mu {\sim} f[S_\kappa]) \cup T_\nu]| = \mu - \kappa + 2\nu\,.$$

Hence, $\lambda \leq \lfloor(\mu - \kappa)/2\rfloor + \nu$ (no matter whether ν is finite or infinite). □

COROLLARY 3.2. *Let $k \in \omega$. A k-generated member of* L *must be isomorphic to $\mathfrak{L}_{\kappa\lambda}$ for some (κ, λ) such that $\kappa + \lambda \leq 2^{k+1} - 1$, and $\kappa \geq 3$ or $\lambda \geq 1$.*

PROOF. Let $k \in \omega$, and suppose that an algebra $\mathfrak{A}$ in L is generated by a set X of k elements. By 3.1(ii), $\mathfrak{A}$ is finite. Therefore, by 3.1(iv),(v)(δ), there is a unique pair (κ, λ) of finite cardinals, with $\kappa \geq 3$ or $\lambda \geq 1$, such that $\mathfrak{A} \cong \mathfrak{L}_{\kappa\lambda}$. Further, by 3.1(ii), $\mathfrak{A}$ is generated by $X \cup \{1'\}$ as a Boolean algebra. As is well known, a Boolean algebra that is generated by $k+1$ elements has at most 2^{k+1} atoms. Since $\mathfrak{L}_{\kappa\lambda}$ has exactly $\kappa + \lambda + 1$ atoms, we conclude that $\kappa + \lambda + 1 \leq 2^{k+1}$. □

LEMMA 3.3. *Let $k \in \omega$, and suppose that ε is an open formula with at most k variables.*

(i) *The following are equivalent.*
   (α) *ε holds in $\mathfrak{L}_{\kappa\lambda}$ for some $\kappa \geq 2^{k+2}$ and some $\lambda \geq 0$.*
   (β) *ε holds in $\mathfrak{L}_{\kappa\lambda}$ for all $\kappa \geq 0$ and all $\lambda \geq 1$.*
   (γ) *ε holds in $\mathfrak{L}_{\kappa 0}$ for all $\kappa \geq 2^{k+1}$ with $\kappa \neq 2$.*

(ii) *Let κ be an arbitrary cardinal. Then the following are equivalent.*
   (α) *ε holds in $\mathfrak{L}_{\kappa\lambda}$ for some $\lambda \geq 2^{k+1}$.*
   (β) *ε holds in $\mathfrak{L}_{\kappa\lambda}$ for all $\lambda \geq 1$.*

PROOF. We begin by drawing a conclusion from 3.1.

(1) Suppose ξ, η, and μ are finite cardinals with $1 \leq \eta$ and $\xi + \eta \leq \mu/2$. Then $\mathfrak{L}_{\xi\eta}$ is embeddable in $\mathfrak{L}_{\mu\nu}$ for every cardinal ν such that $(\mu, \nu) \neq (2, 0)$.

To see this, observe that $\mu \geq 2$, since $1 \leq \eta \leq \mu/2$. If $\nu = 0$, then $\mu \geq 3$, by assumption. Thus, the hypotheses of 3.1(v) are satisfied. Moreover, $\xi \leq \mu$ and

$$\eta \leq \mu/2 - \xi = (\mu - \xi)/2 - \xi/2 \leq (\mu - \xi)/2 + \nu\,.$$

Since η is a natural number, we must have $\eta \leq \lfloor(\mu-\xi)/2\rfloor + \nu$. Applying 3.1(v)($\beta$), we obtain (1).

Turning now to (i), the proofs of the implications from (β) to (α) and from (γ) to (α) are trivial. To prove the reverse implications, from (α) to (β) and (γ), let ε be an open formula with at most k variables. Suppose that (μ, ν) is a pair of cardinals, with $\mu \geq 2^{k+2}$ and $\nu \geq 0$, such that ε is valid in $\mathfrak{L}_{\mu\nu}$. Let (κ, λ) be a pair of cardinals with $\kappa \geq 3$ or $\lambda \geq 1$. To show that ε holds in $\mathfrak{L}_{\kappa\lambda}$, it clearly suffices to show that ε holds in all k-generated subalgebras of $\mathfrak{L}_{\kappa\lambda}$. Let $\mathfrak{A}$ be any such subalgebra. Then by 3.2, $\mathfrak{A}$ is isomorphic to $\mathfrak{L}_{\xi\eta}$ for some pair (ξ, η) of cardinals with $\xi + \eta \leq 2^{k+1} - 1$, and $\xi \geq 3$ or $\eta \geq 1$. We shall show in a moment that, under the assumptions of (β) or (γ), we must have $\eta \geq 1$. From (1) we then obtain that $\mathfrak{L}_{\xi\eta}$ is embeddable into $\mathfrak{L}_{\mu\nu}$. Now ε holds in $\mathfrak{L}_{\mu\nu}$, by assumption. Therefore, ε holds in $\mathfrak{L}_{\xi\eta}$, and hence also in $\mathfrak{A}$.

To prove that $\eta \geq 1$, suppose, for contradiction, that $\eta = 0$. Since

$$\mathfrak{L}_{\xi\eta} \cong \mathfrak{A} \subseteq \mathfrak{L}_{\kappa\lambda}\,,$$

we get $\lambda = 0$ and $\kappa = \xi$, by 3.1(v)(α). But in (β) we have assumed that $\lambda \geq 1$. In (γ) we have assumed that $\kappa \geq 2^{k+1}$, while $\xi \leq 2^{k+1} - 1$; thus, $\xi < \kappa$.

The implication from (β) to (α) in (ii) is also trivial. The proof of the reverse implication is virtually identical to the above proof that (i)(α) imples (i)(β) (taking κ for μ). Instead of appealing to (1) to show that $\mathfrak{L}_{\xi\eta}$ is embeddable into $\mathfrak{L}_{\kappa\nu}$, we argue as follows. Since $\mathfrak{L}_{\xi\eta}$ is embeddable into $\mathfrak{L}_{\kappa\lambda}$, we must have $\xi \leq \kappa$, by 3.1(v). Also,

$$1 \leq \eta \leq 2^{k+1} - 1 < \nu\,.$$

Thus, $\mathfrak{L}_{\xi\eta}$ is embeddable into $\mathfrak{L}_{\kappa\nu}$, by 3.1(v)(β),(γ). $\square$

We now fix the set $\hat{\omega} = \omega{\sim}\{0, 1, 2\}$. For $X \subseteq \hat{\omega}$, the expression ${\sim}X$ will denote the complement of X in $\hat{\omega}$. For each set $X \subseteq \hat{\omega}$, define

$$\mathsf{F}_X = \{\mathfrak{A} : \mathfrak{A} \cong \mathfrak{Ln}(n) \text{ for some } n \text{ in } {\sim}X\}$$

and

$$\mathsf{L}_X = \mathsf{L}{\sim}\mathsf{F}_X\,.$$

In step (4) of the next proof, we shall get a more explicit description of the algebras in L_X.

THEOREM 3.4. (i) *The mapping $X \to \mathsf{L}_X$ is a complete lattice embedding of $\langle Sb\,(\hat{\omega})\,, \cap\,, \cup\rangle$ into the lattice of universal subclasses of L. Its range is the interval $[\mathsf{L}_\varnothing, \mathsf{L}]$ of all universal subclasses of L that include $\mathsf{L}_\varnothing$ (and on this range the complete join and meet operations are just the general union and intersection operations).*

(ii) $\mathcal{E}q(\mathsf{L}_X) = \mathcal{E}q(\mathsf{L}_\varnothing \cup \{\mathfrak{Ln}(n) : n \in X\})$, *but* $\mathcal{E}q(\mathsf{L}_X)$ *is not the equational theory of a finite algebra.*

(iii) $\mathcal{E}q(\mathsf{L}_X)$ *and* X *have the same degree of unsolvability. In particular,* $\mathcal{E}q(\mathsf{L}_X)$ *is recursive iff* X *is recursive.*

(iv) *If* K *is any subclass of* L *that includes* $\mathsf{L}_\varnothing$ *and is closed under isomorphisms, then* K *is in the range of the embedding, i.e.,* $\mathsf{K} = \mathsf{L}_X$ *for some* $X \subseteq \hat{\omega}$.

(v) *If* K *is a universal subclass of* L *that is not in the range of the embedding, then* $\mathcal{E}q(\mathsf{K})$ *is decidable.*

PROOF. For any given integer $n \geq 3$ and any $\mathfrak{A} \in \mathsf{L}$, the algebra $\mathfrak{Ln}(n)$ is embeddable into $\mathfrak{A}$ iff it is isomorphic to $\mathfrak{A}$. Indeed, suppose f is an embedding of $\mathfrak{Ln}(n)$ into $\mathfrak{A}$. By definition of L, $\mathfrak{A}$ is a subalgebra of $\mathfrak{Ln}(\kappa)$ for some $\kappa \geq 3$. Therefore, f is an embedding of $\mathfrak{Ln}(n)$ into $\mathfrak{Ln}(\kappa)$. By 3.1(v)(α), we see that $n = \kappa$ and that f is actually an isomorphism of $\mathfrak{Ln}(n)$ onto $\mathfrak{Ln}(\kappa)$. In particular, $\mathfrak{A} = \mathfrak{Ln}(\kappa)$ and $\mathfrak{Ln}(n) \cong \mathfrak{A}$.

Now let ε_n be the equation that is identically satisfied in a relation algebra $\mathfrak{B}$ iff $\mathfrak{Ln}(n)$ is not embeddable into $\mathfrak{B}$; see Lemma 1.3. As we have just observed, ε_n is identically satisfied in $\mathfrak{A}$ iff $\mathfrak{Ln}(n)$ is not isomorphic to $\mathfrak{A}$. In other words,

(1) $$\mathsf{L}_X = \{\mathfrak{A} \in \mathsf{L} : \mathfrak{A} \models \varepsilon_n \text{ for every } n \text{ in } {\sim}X\}.$$

Since L is a universal class, by 3.1(i), we see from (1) that L_X is a universal subclass of L.

Observe that

(2) $$\mathsf{F}_{\sim X} = \mathsf{F}_\varnothing {\sim} \mathsf{F}_X = \{\mathfrak{A} \in \mathsf{L} : \mathfrak{A} \cong \mathfrak{Ln}(n) \text{ for some } n \text{ in } X\}.$$

With the help of (2) and 3.5(v)(δ), we readily establish

(3) $$\textstyle\bigcup_{i\in I} \mathsf{F}_{\sim X_i} = \mathsf{F}_{\sim \bigcup_{i \in I} X_i} \quad \text{and} \quad \bigcap_{i\in I} \mathsf{F}_{\sim X_i} = \mathsf{F}_{\sim \bigcap_{i \in I} X_i}.$$

For example,

$\mathfrak{A} \in \bigcap_{i\in I} \mathsf{F}_{\sim X_i}$	iff	for each $i \in I$, $\mathfrak{A} \in \mathsf{F}_{\sim X_i}$	
	iff	for each $i \in I$, there is an $n_i \in X_i$ such that $\mathfrak{A} \cong \mathfrak{Ln}(n_i)$,	by (2),
	iff	there is an $m \in \bigcap_{i\in I} X_i$ such that $\mathfrak{A} \cong \mathfrak{Ln}(m)$,	by 3.1(v)(δ),
	iff	$\mathfrak{A} \in \mathsf{F}_{\sim \bigcap_{i \in I} X_i}$	by (2).

(In connection with the third equivalence, recall that when

$$\mathfrak{Ln}(n_i) \cong \mathfrak{A} \cong \mathfrak{Ln}(n_j),$$

we must have $n_i = n_j$, by 3.1(v)(δ).)

By definition, $\mathsf{L}_\varnothing$ consists of all algebras in L that are not isomorphic to some Lyndon algebra on a finite set, i.e., $\mathsf{L}_\varnothing = \mathsf{L} {\sim} \mathsf{F}_\varnothing$. Thus, for each $X \subseteq \hat{\omega}$, we have

$\mathsf{L}_\varnothing \subseteq \mathsf{L}_X$ and

$$\begin{aligned}
\mathsf{L}_X \sim \mathsf{L}_\varnothing &= (\mathsf{L} \sim \mathsf{F}_X) \sim (\mathsf{L} \sim \mathsf{F}_\varnothing) && \text{by definition,}\\
&= \mathsf{L} \cap (\mathsf{F}_\varnothing \sim \mathsf{F}_X) && \text{by Boolean algebra,}\\
&= \mathsf{L} \cap \mathsf{F}_{\sim X} && \text{by (2),}\\
&= \mathsf{F}_{\sim X} && \text{since each } \mathsf{F}_Y \subseteq \mathsf{L}\,.
\end{aligned}$$

In other words,

$$(4) \qquad \mathsf{L}_X = \mathsf{L}_\varnothing \cup \mathsf{F}_{\sim X} = \mathsf{L}_\varnothing \cup \{\mathfrak{A} : \mathfrak{A} \cong \mathfrak{Ln}(n) \text{ for some } n \in X\}\,.$$

Using (3) and (4), we easily check that

$$(5) \qquad \textstyle\bigcap_{i\in I} \mathsf{L}_{X_i} = \mathsf{L}_{\bigcap_{i\in I} X_i} \qquad \text{and} \qquad \bigcup_{i\in I} \mathsf{L}_{X_i} = \mathsf{L}_{\bigcup_{i\in I} X_i}\,.$$

For example,

$$\begin{aligned}
\textstyle\bigcap_{i\in I} \mathsf{L}_{X_i} &= \textstyle\bigcap_{i\in I}(\mathsf{L}_\varnothing \cup \mathsf{F}_{\sim X_i}) && \text{by (4),}\\
&= \mathsf{L}_\varnothing \cup \left(\textstyle\bigcap_{i\in I} \mathsf{F}_{\sim X_i}\right) && \text{by Boolean algebra,}\\
&= \mathsf{L}_\varnothing \cup (\mathsf{F}_{\sim \bigcap_{i\in I} X_i}) && \text{by (3),}\\
&= \mathsf{L}_{\bigcap_{i\in I} X_i} && \text{by (4).}
\end{aligned}$$

Since $X \neq Y$ implies that $\mathsf{F}_X \neq \mathsf{F}_Y$, and hence $\mathsf{L}_X \neq \mathsf{L}_Y$, we see from (5) that the mapping $X \mapsto \mathsf{L}_X$ is a complete lattice embedding.

By (4) we have

$$\mathsf{L} = \mathsf{L}_{\hat\omega} = \mathsf{L}_\varnothing \cup \{\mathfrak{A} : \mathfrak{A} \cong \mathfrak{Ln}(n) \text{ for some } n \in \hat\omega\}\,.$$

Therefore, a class K satisfying $\mathsf{L}_\varnothing \subseteq \mathsf{K} \subseteq \mathsf{L}$ and closed under isomorphisms is completely determined by the Lyndon algebras it contains: if

$$X = \{n \in \hat\omega : \mathfrak{Ln}(n) \in \mathsf{K}\}\,,$$

then $\mathsf{K} = \mathsf{L}_X$. In particular, the lattice embedding maps $\hat\omega$ onto the interval $[\mathsf{L}_\varnothing, \mathsf{L}]$. This completes the proof of (i) and of (iv).

The proof of the first assertion in (ii) follows directly from (4). To prove the second, suppose that $\mathcal{E}q(\mathsf{L}_X)$ is the equational theory of a finite algebra $\mathfrak{B}$. Then the variety $\mathbf{SP}(\mathsf{L}_X)$ is generated by finitely many simple algebras: the subdirectly irreducible factors of $\mathfrak{B}$. Applying Corollary 1.2, we get that $\mathbf{SP}(\mathsf{L}_X)$ has, up to isomorphisms, only finitely many simple algebras. However, L_X is the class of simple algebras in $\mathbf{SP}(\mathsf{L}_X)$, by Theorem 1.1. Certainly L_X has infinitely many pairwise non-isomorphic algebras: we have $\mathsf{L}_\varnothing \subseteq \mathsf{L}_X$, and $\mathfrak{L}_{\kappa 1}$ is in $\mathsf{L}_\varnothing$ for every κ.

Next we take up the proof of (iii). From (4) it follows that

$$\mathcal{E}q(\mathsf{L}_X) = \mathcal{E}q(\mathsf{L}_\varnothing) \cap \mathcal{E}q(\{\mathfrak{Ln}(n) : n \in X\})\,.$$

Therefore, the problem of determining whether an equation ε, say with k variables, is in $\mathcal{E}q(\mathsf{L}_X)$ reduces to the problem of determining whether

(6) $\quad \varepsilon \in \mathcal{E}q(\mathsf{L}_\varnothing)$

and

(7) $\quad \varepsilon \in \mathcal{E}q(\{\mathfrak{Ln}(n) : n \in X\})$.

Of course, (6) holds iff ε is valid in every finitely generated algebra in $\mathsf{L}_\varnothing$. From 3.1(ii),(iv), and the definition of $\mathsf{L}_\varnothing$, we see that

(8) $\quad \mathfrak{A}$ is a finitely generated algebra in $\mathsf{L}_\varnothing$ iff $\mathfrak{A} \cong \mathfrak{L}_{\mu\nu}$ for some finite μ and ν with $\nu \geq 1$.

Thus, applying 3.3(i), we get that (6) holds iff ε is valid in the finite algebra $\mathfrak{L}_{\mu 1}$, where $\mu = 2^{k+2}$. This gives a recursive procedure for determining whether (6) holds: we check the validity of ε in $\mathfrak{L}_{\mu 1}$. If (6) does hold, then ε will automatically be valid in $\mathfrak{Ln}(n)$ for every $n \geq 2^{k+1}$, by 3.3(i). Therefore, to check whether (7) holds, it suffices to determine, for every $n < 2^{k+1}$, whether n is in X, and if so, whether ε is valid in $\mathfrak{Ln}(n)$. This makes it clear that the recursiveness of $\mathcal{E}q(\mathsf{L}_X)$ reduces to that of X. The converse is easy: we can check whether or not a natural number n is in X by checking whether or not ε_n is valid in L_X.

We have shown that the decision problems for $\mathcal{E}q(\mathsf{L}_X)$ and X are reducible to one another, i.e., $\mathcal{E}q(\mathsf{L}_X)$ and X have the same degree of unsolvability. In particular, $\mathcal{E}q(\mathsf{L}_X)$ is recursive iff X is recursive. This completes the proof of (iii).

To prove (v), suppose that K is a universal subclass of L that is not of the form L_X for some $X \subseteq \hat{\omega}$. Since the embedding maps $\hat{\omega}$ onto $[\mathsf{L}_\varnothing, \mathsf{L}]$, by (i), we see that $\mathsf{L}_\varnothing \not\subseteq \mathsf{K}$. If for each finite μ there were a pair (κ, λ) with $\kappa \geq \mu$ such that $\mathfrak{L}_{\kappa\lambda} \in \mathsf{K}$, then we would have $\mathfrak{L}_{\kappa\lambda} \in \mathsf{K}$ for every (κ, λ) with $\lambda \geq 1$, by 3.3(i) and the assumption that K is universal. But then $\mathsf{L}_\varnothing \subseteq \mathsf{K}$, by (8), which is a contradiction. Thus, there must be a finite bound μ_0 on the indices κ such that, for some λ, $\mathfrak{L}_{\kappa\lambda} \in \mathsf{K}$. If for each $\kappa \leq \mu_0$ there is a finite bound on the λ such that $\mathfrak{L}_{\kappa\lambda} \in \mathsf{K}$, then K consist of only finitely many finite algebras. In this case its equational theory is decidable.

Suppose, for some $\kappa \leq \mu_0$ there is no finite bound on the λ such that $\mathfrak{L}_{\kappa\lambda} \in \mathsf{K}$. Let μ_1 be the largest such κ. Then $\mathfrak{L}_{\mu_1\lambda} \in \mathsf{K}$ for every $\lambda \geq 1$, by 3.3(ii) and the assumption that K is universal. Applying 3.1(v)(β) we see that the class

$$\mathsf{M} = \{\mathfrak{L}_{\kappa\lambda} : \kappa \leq \mu_1 \,,\; 1 \leq \lambda\}$$

is included in K. Also, the class

$$\mathsf{N} = \{\mathfrak{L}_{\kappa\lambda} \in \mathsf{K} : \mu_1 < \kappa \leq \mu_0\}$$

is finite, by the maximality of μ_1. Finally, every finitely generated algebra in K is isomorphic either to an algebra in M or to one in N. To check whether an equation ε with k variables is valid in M, it suffices to check whether it holds in $\mathfrak{L}_{\mu_1\nu}$, when $\nu = 2^{k+1}$, by 3.3(ii). Thus, to check whether ε holds in K, we check whether it holds in $\mathfrak{L}_{\mu_1\nu}$ when $\nu = 2^{k+1}$, and we check whether it holds in each of the finitely many finite algebras in N. □

REMARK. For each $X \subseteq \omega$, put $X' = \{n+3 : n \in X\}$. The lattices

$$\langle Sb(\omega), \cap, \cup \rangle \quad \text{and} \quad \langle Sb(\hat{\omega}), \cap, \cup \rangle$$

are obviously isomorphic via the embedding $X \to X'$. Thus, the mapping $X \to \mathsf{L}_{X'}$ is a complete lattice embedding of $\langle Sb(\omega), \cap, \cup \rangle$ onto $\langle [\mathsf{L}_{\varnothing}, \mathsf{L}], \cap, \cup \rangle$ that has properties 3.4(ii)–(v) (provided that we replace "L_X" by "$\mathsf{L}_{X'}$" in (ii)–(iv), "$\mathfrak{Ln}(n)$" by "$\mathfrak{Ln}(n+3)$" in (ii), and "$\hat{\omega}$" by "ω" in (iv)). □

The previous theorem completely describes which classes in the interval $[0, \mathsf{L}]$ of universal subclasses of L have decidable equational theories. We see from this description that there is no smallest class in $[0, \mathsf{L}]$ with an undecidable equational theory. Since the interval $[0, \mathsf{L}]$ is downward closed in the lattice $\mathcal{S}$ of all universal classes of simple relation algebras, we conclude that $\mathcal{S}$ cannot have a smallest member with an undecidable equational theory. In other words, there is no class K in $\mathcal{S}$ with the following properties: K has an undecidable equational theory and, for all M in $\mathcal{S}$, if M has an undecidable equation theory, then $\mathsf{K} \subseteq \mathsf{M}$.

COROLLARY 3.5. *There is no smallest variety of relation algebras with an undecidable equational theory.*

PROBLEM 3.6. Does the lattice of varieties of relation algebras have minimal members with undecidable equational theories? In other words, are there varieties K of relation algebras such that K has an undecidable equational theory, but every proper subvariety of K has a decidable equational theory? □

Problem 3.6 is closely related to Problem 4.18 at the end of Chapter 4.

HISTORICAL REMARKS. Some of the techniques used in this chapter were employed earlier by Jónsson. Lyndon algebras on a projective line were used in Jónsson [1982] to show that there are continuum many varieties of symmetric representable relation algebras, and in Jónsson [1991] to show that any set of equations axiomatizing the variety of all representable relation algebras must contain infinitely many variables. In the first paper Jónsson used the equations ε_n, and he proved that $\mathfrak{Ln}(m)$ is not embeddable into $\mathfrak{Ln}(n)$ when $n \neq m$ (cf. Theorem 3.1(v)(α)). In the second paper Jónsson proved: (1) a subset of the universe of $\mathfrak{Ln}(U)$ is a subuniverse iff it contains 1' and is closed under the Boolean operations — this is essentially just the second assertion of Theorem 3.1(ii); (2) if $\kappa > \lambda$, then a maximal proper subalgebra of $\mathfrak{Ln}(\lambda)$ can always be embedded into $\mathfrak{Ln}(\kappa)$ (cf. Theorem 3.1(v)(β)).

The results of this chapter were obtained by the authors in 1991. They were discussed in lectures given by the second author in 1991 at the Banach Center, but were not announced in print until Andréka-Givant-Németi [1994a].

Németi [1985], page 24, Problem 3, asked whether, for finite dimensional cylindric algebras, there is a smallest variety with an undecidable equational theory. Later, he also posed his question for varieties of relation algebras, and Corollary 3.5 gives the negative answer. □

CHAPTER IV

A FINITELY GENERATED, INFINITE, SIMPLE RELATION ALGEBRA WITH A DECIDABLE EQUATIONAL THEORY

Previously, the only known examples of varieties of relation algebras with decidable equational theories were varieties generated by a single finite algebra or, equivalently, by a finite collection of finite algebras. In the previous chapter we saw examples of varieties with decidable equational theories that are not generated by a finite algebra. In fact, none of the varieties $\mathbf{SP}(\mathsf{L}_X)$ is generated by a single finite algebra. (However, notice for example that $\mathbf{SP}(\mathsf{L}_\varnothing)$ is generated by the single infinite algebra $\mathfrak{Ln}(\omega)$.)

As we saw in Theorem 3.1, each of the varieties $\mathbf{SP}(\mathsf{L}_X)$ is locally finite. The question naturally arises whether every variety of relation algebras — or at least of representable relation algebras — with a decidable equational theory is locally finite. In this chapter we shall construct an example to show that this is not the case. In fact we shall construct a simple, Abelian, representable relation algebra $\mathfrak{A}$ that is infinite but generated by a single element and has a decidable equational theory.

The algebra $\mathfrak{A}$ we shall construct has (for us) a real charm. In particular, it is rather easy to describe. However, the proof that it is representable is not easy, and the proof that its equational theory is decidable is quite involved.

Before describing $\mathfrak{A}$, we introduce some terminology and notation. Let Z denote the set of integers. We shall employ the usual interval notation. For example, if $k, \ell \in Z$, then

$$[k,\ell] = \{n \in Z : k \leq n \leq \ell\} \quad , \quad (k,\ell) = \{n \in Z : k < n < \ell\},$$
$$[k,\infty) = \{n \in Z : k \leq n\} \quad , \quad (-\infty,k) = \{n \in Z : n < k\} \quad .$$

We shall also find it useful to be able to add integers to ∞ and $-\infty$. For each k in Z, we set:

$$k + \infty = \infty \qquad \text{and} \qquad k + -\infty = -\infty\,.$$

The sum $\infty + -\infty$ is not defined.

In the order topology of Z, all intervals are both closed and open. For example, an interval of the form $[k,\ell]$ can be written as $(k{+}{-1}, \ell{+}1)$. Two intervals are of the *same type* provided that they are either both finite, or both have $-\infty$ (respectively, ∞) as a left-hand (respectively, right-hand) endpoint.

The universe of $\mathfrak{A}$ is the collection $Cf(Z)$ of finite and cofinite subsets of Z. The Boolean operations are the usual set-theoretic ones. Thus, $\mathfrak{A}$ is atomic and its atoms are just the elements of the form $\{n\}$ with $n \in Z$. The identity element of $\mathfrak{A}$ is the atom $\{0\}$. Thus, $\mathfrak{A}$ is integral and hence simple. Because $\mathfrak{A}$ is atomic and integral, conversion and relative multiplication are completely determined by specifying their action on subdiversity atoms, i.e., elements of the form $\{n\}$ with $n \in Z\sim\{0\}$. For every $m, n \in Z\sim\{0\}$ we define

$$\{m\}^{\smile} = \{-m\}$$

and

$$\{m\}\,;\{n\} = \begin{cases} [1, m+n] & \text{if } m, n > 0\,, \\ [m+n, -1] & \text{if } m, n < 0\,, \\ (-\infty, -1] \cup [m+n, \infty) & \text{if } m, n \text{ have opposite signs and } \\ & \quad m+n \geq 0, \\ (-\infty, m+n] \cup [1, \infty) & \text{if } m, n \text{ have opposite signs and } \\ & \quad m+n < 0. \end{cases}$$

It is clear from this definition that the relative product of two atoms is either a finite or a cofinite subset of Z. It readily follows that A is closed under relative multiplication.

This seemingly *ad hoc* but quite intriguing definition of relative multiplication is actually dictated by a certain acyclic graph that provides the underlying motivation for, and the key to the representation of, $\mathfrak{A}$. It is not hard to show directly that $\mathfrak{A}$ is a relation algebra. However, we shall not stop to do this now because it will follow directly from the representability of $\mathfrak{A}$. Notice that $\{1\}$ generates $\mathfrak{A}$. In fact,

$$\{1\}^2 = [1,2] \quad , \quad \{1\}^3 = [1,3] \quad , \quad \{1\}^4 = [1,4] \quad , \quad \ldots \quad .$$

Therefore, $\{1\}$ generates the singletons as follows:

$$\{2\} = \{1\}^2\sim\{1\} \quad , \quad \{3\} = \{1\}^3\sim\{1\}^2 \quad , \quad \{4\} = \{1\}^4\sim\{1\}^3 \quad , \ldots$$

and

$$\{-n\} = \{n\}^{\smile}\,.$$

Since the singletons generate all the finite and cofinite sets using only the Boolean operations, we see that $\{1\}$ really does generate $\mathfrak{A}$.

Recall that a (*directed*) *graph* is a structure $\langle U, R\rangle$ where U is a non-empty set and R is a binary relation on U. The graph is *non-trivial* if U has at least two elements. Consider a graph $\langle U, R\rangle$ that satisfies the following three conditions (C1)–(C3).

(C1) R is acyclic, i.e., there is no sequence $z_0, z_1, \ldots, z_n$, with $n > 0$, such that $(z_i, z_{i+1}) \in R$ for each $i < n$, and $z_n = z_0$.

(C2) For any distinct x, y there is a (directed) path from x to y or from y to x, i.e., there is a sequence $z_0, z_1, \ldots, z_n$ such that $(z_i, z_{i+1}) \in R$ for each $i < n$, and either $x = z_0$ and $y = z_n$, or else $y = z_0$ and $x = z_n$.

Before formulating the third condition, we introduce the notion of the distance from one point to another. If there is a directed path from x to y, then all paths between these two points must go from x to y, by (C1), and there will be a path of shortest length, say k. We define the *distance from x to y* to be k, and we write

$$d(x,y) = k \quad \text{and} \quad d(y,x) = -k\,.$$

In particular, $d(x,x) = 0$. The function d is a directed distance function in the following sense. Fix three distinct points x, y, z in U. Since there must be a path between any two of them, and since there are no cycles, there is a unique point among x, y, z (say x) such that there are paths from this point to each of the remaining two points (in this case y and z). There is also a unique point among x, y, z (say y) such there is a path from this point to exactly one of the remaining two points (in this case z). For x and y as specified, we have $d(x,y) + d(y,z) \geq d(x,z)$. (Notice also that $|d(x,y)|$ is a metric.)

(C3) For any x, y with $d(x,y) > 0$ — say $k = d(x,y)$ — and any pair (m,n) of positive integers,

(α) If $m + n > k$, then there is a z such that $d(x,z) = m$ and $d(z,y) = n$ (see Figure 4.1(a));

(β) If $k + m \geq n$, then there is a z such that $d(z,x) = m$ and $d(z,y) = n$ (see Figure 4.1(b));

(γ) If $k + n \geq m$, then there is a z such that $d(x,z) = m$ and $d(y,z) = n$ (see Figure 4.1(c)).

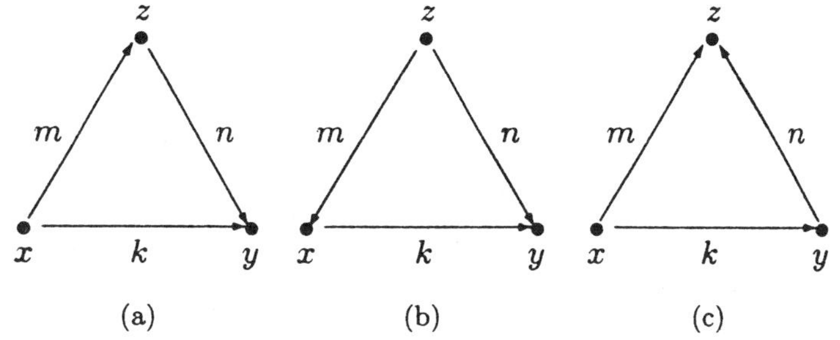

FIGURE 4.1

REMARK. The following consequence of conditions (C1)–(C3) for non-trivial graphs $\langle U, R\rangle$ will be used in the proof of the next lemma: For each x in U and for each positive integer n, we can find elements w and z in A such that $d(x,z) = n$ and $d(w,x) = n$. Indeed, given such an x and n there must be a y in U distinct from x. Set $k = d(x,y)$. If $k > 0$, then $k + n \geq n$; hence, we get z and w by applying (C3)(β),(γ) (with $m = n$). If $k < 0$, we use the same argument with "k" replaced by "$-k$". □

LEMMA 4.1. *Let $\langle U, R\rangle$ be any non-trivial graph satisfying conditions* (C1)–(C3). *Then there is an embedding of $\mathfrak{A}$ into $\mathfrak{R}(U)$ that takes $\{1\}$ to R.*

PROOF. By definition of the distance function, we have:

(1) $d(x, y) = k$ iff $(x, y) \in R^k$ and $(x, y) \notin R^m$ for all m with $0 \leq m < k$, where k is an integer ≥ 0.

(Here, $R^0 = I_U$, the identity relation on U, and $R^{m+1} = R^m|R$.) From conditions (C1) and (C3)(α) we see that if $(x, y) \in R$, then $x \neq y$ and hence $(x, y) \in R^m$ for every $m \geq 1$. In other words, $R \subseteq R^m$ for each m. It follows that

(2) $R^n \subseteq R^m$ whenever $1 \leq n \leq m$.

In view of (1) we obtain

(3) $d(x, y) = k$ iff $(x, y) \in R^k \sim R^{k-1}$, when $k \geq 1$.

Set

(4) $R_k = R^k \sim R^{k-1}$ and $R_{-k} = R_k^{-1}$ for $k \geq 1$, and $R_0 = I_U$.

Then

(5) $R_n \cap R_m = \varnothing$ for distinct $n, m \in Z$.

This follows from (2), (4), and the fact that R is acyclic. In view of (C2), we also have

(6) $\bigcup_{k\in Z} R_k = U \times U$.

We now define a completely additive function f from A to $Sb(U \times U)$. Since f will be entirely determined by specifying its value on singletons, it suffices to stipulate that

$$f(\{m\}) = R_m \quad \text{for} \quad m \in Z .$$

In view of (5) and (6), f is a Boolean embedding of A into $Sb(U \times U)$. From (4) and the definition of $\mathfrak{A}$, it follows that f preserves conversion and the identity element. It remains to show that f preserves relative multiplication between atoms.

We begin by proving that

(7) $R_m | R_n = \bigcup_{i=1}^{m+n} R_i$ for $m, n \geq 1$.

To see this, let $x, y \in U$, and set $k = d(x, y)$. Thus, $(x, y) \in R_k$. If (x, y) is in $R_m|R_n$, then there is certainly a path from x to y of length $m + n$, by (3) and (4). Also, $x \neq y$, since R is acyclic. Therefore we obtain $0 < k \leq m + n$ and hence

$$(x, y) \in \bigcup_{i=1}^{m+n} R_i .$$

For the reverse implication, suppose (x, y) is in $\bigcup_{i=1}^{m+n} R_i$, that is to say, suppose $0 < k \leq m + n$. If $k < m + n$, then by (C3)(α) there is a z such that $d(x, z) = m$ and $d(z, y) = n$. Therefore (x, y) is in $R_m|R_n$. Now suppose that $k = m+n$. Since

$(x, y) \in R_k$, there is a sequence $w_0, \ldots, w_k$ in U such that $(w_i, w_{i+1}) \in R$ for $i < k$ and $x = w_0$, $y = w_k$. Set $z = w_m$. Then $(x, z) \in R_m$; indeed, if there were a path of length less than m from x to z, this would give us a path of length less than $m + n = k$ from x to y. Similarly, $(z, y) \in R_n$. Therefore, (x, y) is in $R_m|R_n$. This proves (7).

(8) $\quad R_m|R_{-n} = R_{-n}|R_m = \bigcup_{i=-1}^{-\infty} R_i \cup \bigcup_{i=m+-n}^{\infty} R_i \quad$ when $\quad m \geq n > 0$.

To prove (8), let $x, y \in U$. Set $k = d(x, y)$ and

$$S = \bigcup_{i=-1}^{-\infty} R_i \cup \bigcup_{i=m+-n}^{\infty} R_i \,.$$

Suppose, first, that (x, y) is in $R_m|R_{-n}$, say $(x, z) \in R_m$ and $(z, y) \in R_{-n}$. We want to show that $(x, y) \in S$. If $k \geq 0$, then there is a path from x to z through y of length $k+n$ (see Figure 4.2(a)). Because $d(x, z) = m$ — see (3) and (4) — we have $m \leq k + n$ and therefore $m + -n \leq k$. Thus, k is one of the indices i in the union defining S, so $(x, y) \in S$. On the other hand, if $k < 0$, then k is automatically one of the indices i defining S, so again $(x, y) \in S$. It follows that $R_m|R_{-n} \subseteq S$. The proof that $R_{-n}|R_m \subseteq S$ is nearly the same. One uses Figure 4.2(b).

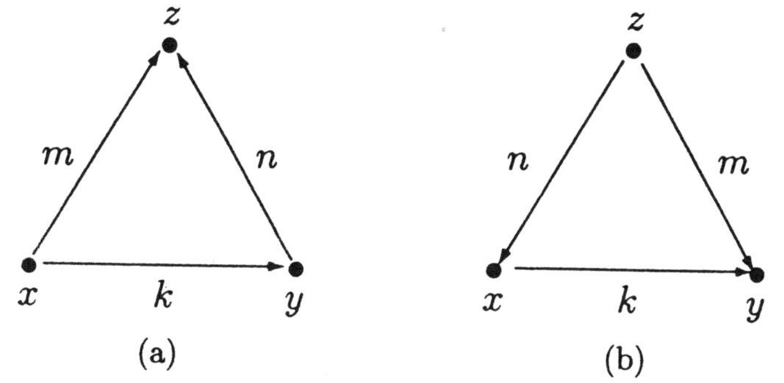

FIGURE 4.2

For the reverse inclusions, suppose that $(x, y) \in S$, with the goal of showing that (x, y) is in $R_m|R_{-n}$. Since $(x, y) \in R_k$, by (3), we see from (5) that k must be one of the indices i defining S; that is, $k \leq -1$ or $k \geq m + -n$. If $k \leq -1$, then since $m \geq n$, we obviously have $-k + m > n$. Apply condition (C3)(γ) to the pair (n, m) — with x and y interchanged, since $d(y, x) = -k > 0$ — to obtain a z such that $d(y, z) = n$ and $d(x, z) = m$ (see Figure 4.3(a)). When $k \geq m + -n$ (≥ 0) we must consider two cases. If $k > 0$, then apply (C3)(γ) directly to obtain a z such that $d(x, z) = m$ and $d(y, z) = n$ (see Figure 4.3(b)). If $k = 0$, then use the remark preceding the lemma to find a z such that $d(x, z) = m$. Since $x = y$ and $m = n$ in this case, we also have $d(y, z) = n$. Thus, in all cases (x, y) is in $R_m|R_{-n}$. This proves that $S \subseteq R_m|R_{-n}$.

To prove that $S \subseteq R_{-n}|R_m$, we argue in very much the same way. When $k \leq -1$, we use condition (C3)(β) — with x and y interchanged — to obtain a z such that $d(z, y) = m$ and $d(z, x) = n$ (see Figure 4.4(a)). When $k \geq m + -n$ (≥ 0), we

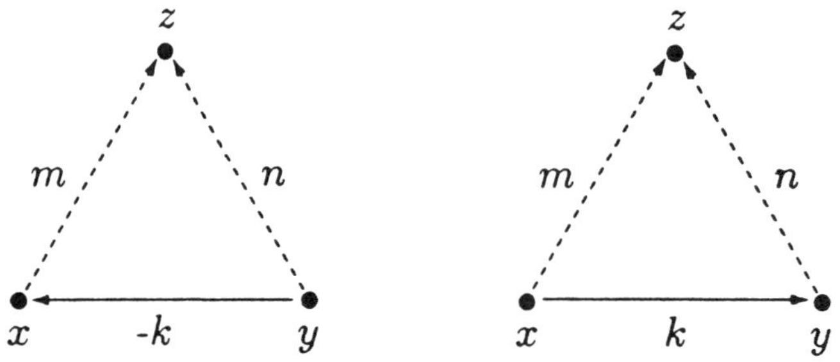

(a) (b)

FIGURE 4.3

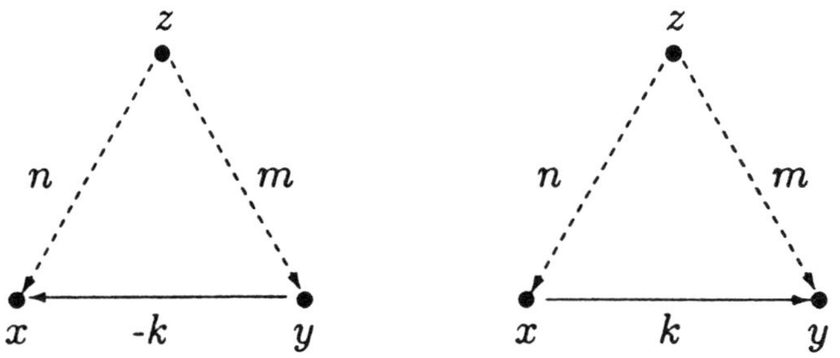

(a) (b)

FIGURE 4.4

again use either (C3)(β) applied to the pair (n, m) or else the remark preceding the lemma to obtain a z such that $d(z, x) = n$ and $d(z, y) = m$ (see Figure 4.4(b)). This completes the proof of (8).

By applying $\smile$ to both sides of (7) and (8) we get

(9) $\quad R_m | R_n = \bigcup_{i=-1}^{m+n} R_i \quad$ for $\quad m, n \leq -1\,,$

(10) $\quad R_m | R_{-n} = R_{-n} | R_m = \bigcup_{i=1}^{\infty} R_i \cup \bigcup_{i=m+-n}^{-\infty} R_i \quad$ when $\quad n \geq m > 0\,.$

Comparing (7)–(10) with the definition of relative multiplication between atoms of $\mathfrak{A}$, we see that f preserves the relative multiplication of atoms. Since f is completely additive and relative multiplication is completely distributive, it follows that f preserves arbitrary relative products. This completes the proof of the lemma. $\square$

THEOREM 4.2. *$\mathfrak{A}$ is representable.*

PROOF. By Lemma 4.1, it suffices to construct a non-trivial graph $\mathfrak{U} = \langle U, R \rangle$ satisfying conditions (C1)–(C3). To construct $\mathfrak{U}$ we shall build a chain of finite, acyclic graphs $\mathfrak{U}_j = \langle U_j, R_j \rangle$ for $j < \omega$, such that $\mathfrak{U}_{j+1}$ witnesses the verification of conditions (C2) and (C3) in $\mathfrak{U}_j$ up to paths of length j. We shall then take $\mathfrak{U} = \bigcup_{j<\omega} \mathfrak{U}_j$. Here is an outline of the construction.

Fix two arbitrary, distinct elements, take U_0 to be the set of these two elements, and take R_0 to be the empty relation. Suppose, now, that $\mathfrak{U}_j$ has been defined.

If two points $x, y \in U_j$ are connected by a path in $\mathfrak{U}_j$, we shall say that x and y are *j-connected*. In this case we denote the length of the minimal path between them in $\mathfrak{U}_j$ by $|x,y|_j$. Set $\ell = |U_j|$.

(1) Put the elements of U_j into U_{j+1}, and the pairs of R_j into R_{j+1}.

(2) For distinct $x, y \in U_j$ that are not j-connected, add $\ell - 1$ new points to U_{j+1} and put into R_{j+1} a new path of length ℓ from just one of the points x, y to the other — say, from x to y — i.e., add new points $v_0, \ldots, v_{\ell-2}$ to U_{j+1} and pairs (x, v_0), $(v_{\ell-2}, y)$, and (v_i, v_{i+1}) for $i < \ell - 2$, to R_{j+1}; see Figure 4.5. (This takes care of condition (C2) for $\mathfrak{U}_j$. In fact, ℓ will be the distance from x to y in the final graph.)

(3) For distinct $x, y \in U_j$ that are j-connected — say the path runs from x to y — and for any pair (m, n) satisfying $0 < m, n \leq j$:

   (a) if $m + n > |x,y|_j$, add a new path (consisting of all new points) of length $m + n$ from x to y;

   (b) if $|x,y|_j + m \geq n$, add a new point z, a new path of length m from z to x, and a new path of length n from z to y;

   (c) if $|x,y|_j + n \geq m$, add a new point w, a new path of length m from x to w, and a new path of length n from y to w.

   (This takes care of the conditions in (C3).)

FIGURE 4.5

Since $\mathfrak{U}_j$ is finite, and since each new path we add (for given x and y in U_k) is of length $\leq 2\ell$ (since $j < |U_j| = \ell$), it is clear that the graph $\mathfrak{U}_{j+1}$ will be finite.

Caution must be exercised in carrying out the construction in step (2). For example, it may happen that no one of some three points x, y, z in U_j is j-connected to the other two points, and in constructing paths in $\mathfrak{U}_{j+1}$, we adjoin a path from x to y, a path from y to z, and a path from z to x , thus creating a cycle. More complicated scenarios involving more points from U_j can also be imagined. To avoid this difficulty, we define a *scheduling function* g_j, i.e., a one-one function on some natural number that enumerates those unordered pairs $\{x, y\}$ of distinct points in U_j that are not j-connected. We then carry out step (2) inductively, according to g_j. If $g_j(p) = \{x, y\}$, then in choosing whether to construct a path from x to y or vice versa, we first check whether any of the paths adjoined in steps 0 through $p-1$ have created a path between x and y. If so, then we add the new path of length ℓ (according to the dictates of (2)) in the same direction as the existing path. (An inductive argument is needed to show that there cannot be two such existing paths that go in opposite directions.) If no such paths have yet been created, then we are free to choose whether the path goes from x to y or from y to x.

Similarly, we can use a scheduling function h_j to carry out part (3) of the construction (say, after part (2) is done). The function h_j enumerates the pairs $(\{x, y\},\ (m, n))$ such that x, y are distinct elements of U_j that are j-connected, and $0 < m, n \leq j$. If $h_j(p) = (\{x, y\},\ (m, n))$, then there may be one, two, or three substeps here, according to which cases of (a)–(c) apply. (For instance, when $|x, y|_j = 3$, $m = 4$ and $n = 5$, all three cases apply.) It is important that at each step (and substep) of the construction the new elements we add are completely different from elements added in all previous steps.

Fix, now, distinct $x, y \in U_j$.

(4) Suppose there is a path of length p in $\mathfrak{U}_{j+1}$ connecting x and y, and consisting entirely of new points (except for x and y). If x and y are not j-connected, then $p = \ell$. If x and y are j-connected, then $p > |x, y|_j$.

Indeed, since x, y are distinct and since the path consists entirely of new points, it must have been created when the pair $\{x, y\}$ was scheduled, either in step (2) or in step (3). If x and y are not j-connected, then the path was created in step (2). In this case $p = \ell$, since we added a path of length ℓ between x and y when $\{x, y\}$ was scheduled. If x and y are j-connected, then the path was created during step (3). It must have been created during substep (a), since substeps (b) and (c) do not lead to paths between x and y. In substep (a) we added a path of length $m+n$ when $m + n > |x, y|_j$; Thus, $p > |x, y|_j$.

(5) If x and y are j-connected, then $|x, y|_{j+1} = |x, y|_j$. If they are not j-connected, then $|x, y|_{j+1} = \ell$.

To prove (6), let $z_0, z_1, \ldots, z_p$ be a path between x and y in $\mathfrak{U}_{j+1}$ of the minimal length $p = |x, y|_{j+1}$. Thus, $\{x, y\} = \{z_0, z_p\}$. If x and y are j-connected, then obviously $p \leq |x, y|_j$, and if they are not j-connected, then obviously $p \leq \ell$ (since we added a path of length ℓ in $\mathfrak{U}_{j+1}$). Thus, our goal is to show that $p \geq |x, y|_j$ in the former case, and $p \geq \ell$ in the latter case.

Suppose that the entire path $z_0, \ldots, z_p$ is in $\mathfrak{U}_j$. Then x and y are certainly j-connected. Obviously, in this case $p \geq |x, y|_j$, by the minimality of $|x, y|_j$.

We may assume, therefore, that at least one point in the path is not in $\mathfrak{U}_j$. Thus, we can find integers s and t in $\{0, \ldots, p\}$ such that $s + 1 < t$ (i.e., the path from z_s to z_t has length at least 2), $z_s, z_t \in U_j$, and $z_r \notin U_j$ for $s < r < t$. (If z_q is a point of the path not in U_j, let s be the largest integer below q such that $z_s \in U_j$; s certainly exists, since $z_0 \in U_j$. Similarly, let t be the smallest integer above q such that $z_t \in U_j$; t exists, since $z_p \in U_j$.) Thus, z_s and z_t are distinct points in $\mathfrak{U}_j$ and the path between them consists entirely of new points. If z_s and z_t are j-connected, then the length $t - s$ of the path between them is greater than $|z_s, z_t|_j$, by (4). But then we can replace that segment of the given path between z_s and z_t by a smaller path, thus constructing a smaller path between x and y. This contradicts the minimality of p. Hence, z_s and z_t cannot be j-connected. Applying (4) again, we conclude that $t - s = \ell$. Thus, $p \geq \ell$.

If x and y are not j-connected, we are done. If they are j-connected (by some other path), then obviously $|x, y|_j \leq |U_j| = \ell \leq p$. (Here we use for the first and only time the fact that the length of the new paths we add is actually $|U_j|$.) This proves (5).

From (5) we conclude:

(6) If j is the smallest integer such that x and y are j-connected, then for all $k > j$ we have $|x,y|_k = |x,y|_j$.

Next, we establish:

(7) $\mathfrak{U}_j$ is acyclic.

The proof of (7) is by induction on j. For $j = 0$ it is obvious. Suppose, now, that $\mathfrak{U}_j$ is acyclic. In constructing $\mathfrak{U}_{j+1}$, no cycle arises in step (1), by the induction hypothesis. To show that no cycle is created during step (2) of the construction we proceed by strong induction on the integers in the domain of the scheduling function g_j. Let p be in the domain of g_j, with $g_j(p) = \{x,y\}$, and suppose that prior to carrying out this stage of step (2) no cycle has been created. (This is certainly true when $p = 0$, by our previous remarks.) The only way a cycle could arise at stage p is if we construct, say, a path from x to y when there already exists a path from y to x. But in this case *all* prior paths between x and y go from y to x, since we are assuming no cycles exist prior to carrying out stage p. In this case, we explicitly have required that the path created at stage p also goes from y to x. Thus, no cycle can be created at stage p.

The proof that no cycle can be created during step (3) of the construction is similar, and is left to the reader. Notice that only substep (a) need be considered. In substeps (b) and (c) we do not create a path between x and y at all, but rather only paths to or from some new point. This completes the proof of (7).

Let $\mathfrak{U}$ be the union of the graphs $\mathfrak{U}_j$ for $j < \omega$. The non-triviality of $\mathfrak{U}$ follows from the fact that U_0 has two elements. Since the union of a chain of acyclic graphs is acyclic, $\mathfrak{U}$ satisfies condition (C1). Now fix distinct x, y in $\mathfrak{U}$. Then x, y are in $\mathfrak{U}_j$ for some smallest j. If x, y are not already j-connected, then in carrying out step (2) in the construction of $\mathfrak{U}_{j+1}$ they become $(j+1)$-connected. Hence, $\mathfrak{U}$ satisfies condition (C2). Assume that the path between x and y goes from x to y. By (6) we have $d(x,y) = |x,y|_j$ or $d(x,y) = |x,y|_{j+1}$. Given a pair (m,n) of positive integers, set $p = \max\{m,n,j+1\}$. Then in step (3) of the construction of $\mathfrak{U}_{p+1}$ we create paths satisfying the conditions (α)–(γ) of (C3). This completes the proof of the theorem. □

We come now to the decision procedure for $\mathcal{E}q(\mathfrak{A})$. The possibility of such a procedure for the equational theory of a relation algebra appears to depend upon the complexity of the algebra's arithmetic and, above all, upon the complexity of the relative multiplication operation. In the case of $\mathfrak{A}$, the relative product of two sets, x_1 and x_2, is really a function of five specific attributes of each set; it does not depend at all on the general contents of the two sets. It is this simplicity in the arithmetic table of relative multiplication that makes a decision procedure for $\mathfrak{A}$ possible.

The five attributes of sets that must be known to compute the relative product of two sets are indicated by the following five functions from A to Z.

DEFINITION 4.3. We define *indicator functions*

$$P \quad , \quad p \quad , \quad N \quad , \quad n \quad , \quad z$$

from $A = Cf(Z)$ to Z. For each $x \in A$, the values

$$P(x) \quad , \quad p(x) \quad , \quad N(x) \quad , \quad n(x) \quad , \quad z(x)$$

are called the *indicators* of x.

$$P(x) = \begin{cases} \sup\{k \in x : 1 \leq k\} & \text{if } \{k \in x : 1 \leq k\} \neq \varnothing, \\ 0 & \text{otherwise.} \end{cases}$$

$$p(x) = \begin{cases} \inf\{k \in x : 1 \leq k\} & \text{if } \{k \in x : 1 \leq k\} \neq \varnothing, \\ 0 & \text{otherwise.} \end{cases}$$

$$N(x) = \begin{cases} \inf\{k \in x : k \leq -1\} & \text{if } \{k \in x : k \leq -1\} \neq \varnothing, \\ 0 & \text{otherwise.} \end{cases}$$

$$n(x) = \begin{cases} \sup\{k \in x : k \leq -1\} & \text{if } \{k \in x : k \leq -1\} \neq \varnothing, \\ 0 & \text{otherwise.} \end{cases}$$

$$z(x) = \begin{cases} 1 & \text{if } 0 \in x, \\ 0 & \text{otherwise.} \end{cases}$$

The set

$$In_x = \{P(x)\,,\, p(x)\,,\, N(x)\,,\, n(x)\,,\, z(x)\}$$

is called the *set of indicators of* x. □

Using these indicators, it is easy to describe the arithmetic of relative multiplication. We omit those rules that follow by the commutativity of ; .

LEMMA 4.4 (ARITHMETIC RULES FOR RELATIVE MULTIPLICATION). *Let* x_1, x_2 *be elements of* A *that are different from* 0 *and* 1'. *For* $i = 1, 2$, *set*

$$P_i = P(x_i) \quad , \quad p_i = p(x_i) \quad , \quad N_i = N(x_i) \quad , \quad n_i = n(x_i) \quad , \quad z_i = z(x_i)\,.$$

(i) *If each of* p_1, n_1, p_2, n_2 *is different from* 0, *then*

$$x_1 \,;\, x_2 = \begin{cases} 0' & \text{if } x_1 \cap x_2^{\smile} = 0, \\ 1 & \text{if } x_1 \cap x_2^{\smile} \neq 0. \end{cases}$$

(ii) *If* $p_2, n_2 \neq 0$ *and* $n_1 = 0$, *then*

$$x_1 \,;\, x_2 = \begin{cases} (-\infty, P_1 + n_2] \cup [1, \infty) & \text{if } P_1 + n_2 < 0, \\ 0' & \text{if } P_1 + n_2 \geq 0 \text{ and } \\ & x_1 \cap x_2^{\smile} = 0, \\ 1 & \text{if } P_1 + n_2 \geq 0 \text{ and } \\ & x_1 \cap x_2^{\smile} \neq 0. \end{cases}$$

(iii) *If* $p_2, n_2 \neq 0$ *and* $p_1 = 0$, *then*

$$x_1 \,;\, x_2 = \begin{cases} (-\infty, -1] \cup [N_1 + p_2, \infty) & \text{if } N_1 + p_2 > 0, \\ 0' & \text{if } N_1 + p_2 \leq 0 \text{ and } x_1 \cap x_2^{\smile} = 0, \\ 1 & \text{if } N_1 + p_2 \leq 0 \text{ and } x_1 \cap x_2^{\smile} \neq 0. \end{cases}$$

(iv) *If* $p_1 = 0$ *and* $n_2 = 0$, *then*

$$x_1 \,;\, x_2 = \begin{cases} (-\infty, -1] \cup [N_1 + p_2, \infty) & \text{if } N_1 + p_2 > 0, \\ (-\infty, n_1 + P_2] \cup [1, \infty) & \text{if } n_1 + P_2 < 0, \\ 0' & \text{if } N_1 + p_2 \leq 0 \leq n_1 + P_2 \text{ and } x_1 \cap x_2^{\smile} = 0, \\ 1 & \text{if } N_1 + p_2 \leq 0 \leq n_1 + P_2 \text{ and } x_1 \cap x_2^{\smile} \neq 0. \end{cases}$$

(v) *If* $n_1 = n_2 = 0$, *then*

$$x_1 \,;\, x_2 = \begin{cases} [1, P_1 + P_2 + 1) & \text{if } z_1 = 0 \text{ or } z_2 = 0, \\ [0, P_1 + P_2 + 1) & \text{if } z_1 = z_2 = 1. \end{cases}$$

(vi) *If* $p_1 = p_2 = 0$, *then*

$$x_1 \,;\, x_2 = \begin{cases} (N_1 + N_2 + -1, -1] & \text{if } z_1 = 0 \text{ or } z_2 = 0, \\ (N_1 + N_2 + -1, 0] & \text{if } z_1 = z_2 = 1. \end{cases}$$

REMARKS. Some remarks regarding the notation used in the formulation of the lemma might serve to clarify its meaning. To say that $n_i = 0$ or, equivalently, that $N_i = 0$ is to say that x_i has no negative numbers; to say that $p_i = 0$ or, equivalently, that $P_i = 0$ is to say that x_i has no positive numbers. Therefore, to assume, for example, that $p_2, n_2 \neq 0$ is to assume that x_2 contains both positive and negative numbers. The stipulation that x_i be different from 0 and 1' reduces to the stipulation that x_i contain a non-zero integer. Thus, we are requiring that either n_i , $N_i \neq 0$ or p_i , $P_i \neq 0$. We always have $n_i \neq -\infty$ and $p_i \neq \infty$.

Notice that 0 is being used in two different ways in the lemma. For example, in its first occurrence in (i) it denotes the integer zero; in its second and third occurrence in (i) it denotes the Boolean zero of the relation algebra $\mathfrak{A}$. This should not cause confusion. The notation $[1, P_1 + P_2 + 1)$ is being used to handle the case when one of P_1 and P_2 is ∞. When both are finite, then $[1, P_1 + P_2 + 1)$ is the same as $[1, P_1 + P_2]$. However, when at least one of P_1 and P_2 is ∞, then $[1, P_1 + P_2 + 1)$ is really $[1, \infty)$. Notice that, in this case, the notation $[1, P_1 + P_2]$ does not make sense, since it denotes $[1, \infty]$. Similarly, the notation $(N_1 + N_2 + -1, -1]$ is used to handle the case when one of N_1 and N_2 is $-\infty$. When both are finite, then $(N_1 + N_2 + -1, -1]$ is the same as $[N_1 + N_2, -1]$. When at least one of them is $-\infty$, then $(N_1 + N_2 + -1, -1]$ is $(-\infty, -1]$. □

PROOF. Notice, first of all, that

(1) $1' \leq x_1 \,;\, x_2$ iff $x_1 \cap x_2^{\smile} \neq 0$.

Indeed,

$$\begin{aligned} 1' \leq x_1 \,;\, x_2 \quad &\text{iff} \quad (x_1 \,;\, x_2) \cap 1' \neq 0 && \text{since 1' is an atom,} \\ &\text{iff} \quad x_1 \cap (1' \,;\, x_2{}^{\smile}) \neq 0 && \text{by the cycle laws,} \\ &\text{iff} \quad x_1 \cap x_2{}^{\smile} \neq 0 && \text{since 1' is the identity element.} \end{aligned}$$

This proves (1).

(2) If $n_1, P_2 \neq 0$, then

$$\begin{aligned} (-\infty, -1] \cup (n_1 + P_2 + -1, \infty) &\subseteq x_1 \,;\, x_2 && \text{when } n_1 + P_2 \geq 0\,, \\ (-\infty, n_1 + P_2] \cup [1, \infty) &\subseteq x_1 \,;\, x_2 && \text{when } n_1 + P_2 < 0. \end{aligned}$$

To prove (2), suppose that n_1, $P_2 \neq 0$, i.e., suppose that x_1 contains negative numbers and x_2 positive numbers. The inequality $n_1 + P_2 \geq 0$ is equivalent to the assertion that, for some negative $m \in x_1$ and some positive $q \in x_2$, we have $m + q \geq 0$. For any such m and q, we certainly have $m + q \leq n_1 + P_2$; therefore, since

$$\{m\} \,;\, \{q\} = (-\infty, -1] \cup [m + q, \infty)$$

by the relative multiplication rules for $\mathfrak{A}$, we see that $\{m\} \,;\, \{q\}$ includes

$$(-\infty, -1] \cup (n_1 + P_2 + -1, \infty)\,,$$

and is included in $x_1 \,;\, x_2$. This gives us the first inclusion in (2). (See the remarks preceding the proof for an explanation of why we write "$(n_1 + P_2 + -1, \infty)$" instead of "$[n_1 + P_2, \infty)$".) Suppose, now, that $n_1 + P_2 < 0$, i.e., for all negative $m \in x_1$ and all positive $q \in x_2$, we have $m + q < 0$. Certainly, $n_1 \neq -\infty$. Also, $P_2 \neq \infty$, since

$$n_1 + P_2 < 0 < n_1 + \infty = \infty\,.$$

Because n_1 and P_2 are non-zero, it follows that $n_1 \in x_1$ and $P_2 \in x_2$. Thus, $\{n_1\} \,;\, \{P_2\}$ is included in $x_1 \,;\, x_2$. This and the relative multiplication rules for $\mathfrak{A}$ give the second inclusion in (2).

Each of the following can be proved in a similar fashion; we leave the details of the verification to the reader. (Alternately, using the equality $x_1 \,;\, x_2 = x_2 \,;\, x_1$ we can derive (3) from (2) . Further, using the equalities

$$(x_1 \,;\, x_2)^{\smile} = x_1{}^{\smile} \,;\, x_2{}^{\smile} \quad , \quad -N(x_i) = P(x_i{}^{\smile}) \quad , \quad -n(x_i) = p(x_i{}^{\smile})$$

we can derive (4) from (3), and (6) from (5).)

(3) If $P_1, n_2 \neq 0$, then

$$(-\infty, -1] \cup (P_1 + n_2 + -1, \infty) \subseteq x_1 \,;\, x_2 \qquad \text{when } P_1 + n_2 \geq 0,$$
$$(-\infty, P_1 + n_2] \cup [1, \infty) \subseteq x_1 \,;\, x_2 \qquad \text{when } P_1 + n_2 < 0.$$

(4) If $N_1, p_2 \neq 0$, then

$$(-\infty, N_1 + p_2 + 1) \cup [1, \infty) \subseteq x_1 \,;\, x_2 \qquad \text{when } N_1 + p_2 \leq 0,$$
$$(-\infty, -1] \cup [N_1 + p_2, \infty) \subseteq x_1 \,;\, x_2 \qquad \text{when } N_1 + p_2 > 0.$$

(5) If $N_1, N_2 \neq 0$, then

$$(N_1 + N_2 + -1, -1] \subseteq x_1 \,;\, x_2 \,.$$

(6) If $P_1, P_2 \neq 0$, then

$$[1, P_1 + P_2 + 1) \subseteq x_1 \,;\, x_2 \,.$$

Assume, now, that the hypothesis of (i) holds. Then each of P_1, N_1, P_2, N_2 is also different from 0. If $n_1 + P_2 \geq 0$, then $x_1 \,;\, x_2$ includes the intervals

$$(-\infty, -1] \quad , \quad (n_1 + P_2 + -1, \infty) \quad , \quad [1, P_1 + P_2 + 1)\,,$$

by (2) and (6). If $n_1 + P_2 < 0$, then $x_1 \,;\, x_2$ includes the intervals

$$(-\infty, n_1 + P_2] \quad , \quad [1, \infty) \quad , \quad (N_1 + N_2 + -1, 1]\,,$$

by (2) and (5). In both cases, we see that $0' \leq x_1 \,;\, x_2$, i.e.,

$$(-\infty, -1] \cup [1, \infty) \subseteq x_1 \,;\, x_2 \,.$$

Part (i) follows from this and (1).

Turning to (ii), the assumption that $n_1 = 0$ and $x_1 \neq 0, 1'$ implies that $P_1 \neq 0$. (See the remarks preceding the proof of the lemma). When $P_1 + n_2 \geq 0$, we have $0' \leq x_1 \,;\, x_2$, just as in the proof of (i). Suppose that $P_1 + n_2 < 0$. Then

$$(-\infty, P_1 + n_2] \cup [1, \infty) \subseteq x_1 \,;\, x_2 \,,$$

by (3). Clearly, the sum of every positive number q from x_1 with every negative number m from x_2 is less than or equal to $P_1 + n_2$, by the definitions of P_1 and n_2; in particular, such a sum is less than 0, so the relative product

$$\{q\} \,;\, \{m\} = (-\infty, q + m] \cup [1, \infty)$$

is included in

$$(-\infty, P_1 + n_2] \cup [1, \infty)\,.$$

Since x_1 is assumed to contain no negative numbers, it follows that $x_1 ; x_2$ can contain no negative numbers greater than $P_1 + n_2$. Together with (1), this proves (ii).

The proof of (iii) is similar to that of (ii), but uses (4) instead of (3); alternately, (iii) can be derived from (ii) in the same way that (4) can be derived from (3) (see the remarks preceding (3)). The proofs of (v) and (vi) are easy and use (5) and (6). Assume, now, that the hypothesis of (iv) holds. Suppose, first, that $N_1 + p_2 > 0$. Then

$$(-\infty, -1] \cup [N_1 + p_2, \infty) \subseteq x_1 ; x_2 ,$$

by (4). Moreover, in this case the sum of every negative number in x_1 with every positive number in x_2 is greater than or equal to $N_1 + p_2$; in particular, such a sum is greater than 0. Therefore, the relative product of two such numbers is included in

$$(-\infty, -1] \cup [N_1 + p_2, \infty) .$$

Thus, $x_1 ; x_2$ can contain no positive numbers below $N_1 + p_2$. The case when $n_1 + P_2 < 0$ is treated similarly, and the other cases are argued as in (i) and (ii). This completes the proof. □

An equation $\tau = \sigma$ is valid in $\mathfrak{A}$ iff the equation $-\tau \cdot \sigma + \tau \cdot -\sigma = 0$ is valid, iff the inequality $-\tau \cdot \sigma + \tau \cdot -\sigma \neq 0$ is not satisfied. Thus, to give a decision procedure for the equational theory of $\mathfrak{A}$, we only need give a procedure that determines, for each term τ, whether there is an assignment a from $\mathfrak{A}$ such that $\tau[a] \neq \varnothing$.

At first blush, it might seem that such a procedure would involve the computation of the possible values of $\tau[a]$ in $\mathfrak{A}$, e.g., by bringing each term τ into some normal form. However, we have seen that the relative product of two sets in $\mathfrak{A}$ does not depend on the general contents of these sets, but rather only on five key elements: the indicators of the sets. Thus, in some sense we can replace computations in $\mathfrak{A}$ — which is an algebra of *sets of integers* — by computations in algebras of *integers* (together with $\pm\infty$). This will considerably simplify our task.

Even more is true. It is not the actual value of the relevant indicators that is important, but only such properties as their order among one another, their relationship to 0, and whether they are "close" to one another or "far apart". Thus, if there is an assignment a from $\mathfrak{A}$ such that $\tau[a] \neq \varnothing$, then it is possible to construct another assignment a' satisfying $\tau[a'] \neq \varnothing$, and such that the indicators of $\tau[a']$ (except for $\pm\infty$) come from an interval whose endpoints are computable from τ. Thus, roughly speaking, to decide if an inequality $\tau \neq 0$ is satisfiable, we first construct (using finite algebras of extended integers) a certain finite interval. We then compute all of the (finitely many) assignments that can arise from this interval and check them one-by-one to see if they satisfy $\tau \neq 0$.

We define Z_e to be the set of *extended integers*:

$$Z_e = Z \cup \{-\infty, \infty\} ,$$

where $-\infty$ and ∞ are two distinct elements that do not occur in Z. We define $<$ on Z_e in the usual way. Thus, e.g., $-\infty < m < \infty$ for every $m \in Z$. Let $+$ be the ternary relation of addition on Z_e; in other words, the relation $+(k, \ell, m)$ holds just in case m is the sum of k and ℓ. We shall continue to write $k + \ell = m$, but it is important to keep in mind that $+$ is a relation, not an operation. Finally, $-$ is the unary operation of forming negatives. In particular, $-(\infty) = -\infty$ and $-(-\infty) = \infty$. Set

$$\mathfrak{Z}_e = \langle Z_e\,,\, +\,,\, -\,,\, <\,,\, 0\,,\, 1\,,\, \infty\rangle\,.$$

A structure

$$\mathfrak{M} = \langle M\,,\, +\,,\, -\,,\, <\,,\, 0\,,\, 1\,,\, \infty\rangle$$

is a *substructure* of $\mathfrak{Z}_e$ if M is a subset of Z_e, the operations and relations of $\mathfrak{M}$ are just the restrictions of the corresponding operations and relations of $\mathfrak{Z}_e$ to M, and the constants of $\mathfrak{M}$ are equal to the corresponding constants of $\mathfrak{Z}_e$.

Suppose that $M \subseteq Z_e$ and that I is an interval of Z. Then I is a *maximal interval in* M if I is a subset of M, but no interval of Z properly including I is a subset of M. Also, if $I = (k, \ell)$, then I is a *gap* in M if k and ℓ are in M, but $I \cap M = \varnothing$.

In our subsequent development we shall refer several times to the *endpoints* of an interval I in $\mathfrak{Z}$. Contrary to common usage, we do not mean the smallest and largest numbers in the interval; such numbers may in fact not exist at all. Rather, we imagine an implicit or explicit representation of I in interval notation — say $I = (k, \ell)$ or $I = [k, \ell]$ — and we intend the endpoints of I to be k and ℓ; more exactly, k is the *lefthand endpoint* and ℓ the *righthand endpoint* of I. For example, the lefthand and righthand endpoints of $(-\infty, \infty)$ are $-\infty$ and ∞ respectively; the endpoints of $(-3, 6]$ are -3 and 6. Notice that the notion of the endpoints of an interval is not an intrinsic property of the interval, but rather depends on the interval's representation. For example,

$$(-8, 6] \quad , \quad [-7, 7) \quad , \quad (-8, 7) \quad , \quad [-7, 6]$$

all represent the same interval of $\mathfrak{Z}$, but have different endpoints.

As mentioned above, isomorphic finite substructures of $\mathfrak{Z}_e$ will play an important role in our decision procedure. Before introducing the somewhat involved terminology, notation, and lemmas needed to establish the connection between these substructures and the satisfaction of inequalities in $\mathfrak{A}$, we develop some basic properties of these substructures. The reader should recall, from the introductory remarks of this chapter, the notion of two intervals of Z being of the same type.

LEMMA 4.5. *Suppose $\mathfrak{M}$ is a substructure of $\mathfrak{Z}_e$.*

(i) *Every non-empty interval I of Z with endpoints in M has a unique decomposition over M into the disjoint union of maximal non-empty intervals in $I \cap M$ and non-empty gaps in M.*

(ii) *Suppose f is an isomorphism from $\mathfrak{M}$ to another substructure $\mathfrak{M}'$ of $\mathfrak{Z}_e$.*

  (α) *If (k, ℓ) is a gap in $\mathfrak{M}$, then $(f(k), f(\ell))$ is a gap in $\mathfrak{M}'$.*

  (β) *For every interval I of Z that is included in M, the image $f[I]$ is an interval in $\mathfrak{M}'$ of the same type. Moreover, I is maximal in $\mathfrak{M}$ iff $f[I]$ is maximal in $\mathfrak{M}'$.*

(γ) *Let I be an interval of Z with endpoints k, ℓ in M, and I' the interval of the same type with endpoints $f(k)$ and $f(\ell)$. Then the correspondence between gaps given in* (α), *the correspondence between intervals given in* (β), *maps the unique decomposition of I over M given in* (i) *to the unique decomposition of I' over M'.*

PROOF. Fix a non-empty interval I of Z. The union of a chain of intervals in $I \cap M$ is again an interval in $I \cap M$. Thus, by Zorn's Lemma, every element of $I \cap M$ is contained in a maximal interval in $I \cap M$. If two intervals in $I \cap M$ have a non-empty intersection, then their union is an interval in $I \cap M$. Therefore, two distinct maximal intervals in $I \cap M$ are disjoint. The collection $\mathcal{S}$ of maximal non-empty intervals included in $I \cap M$ is, of course, uniquely determined. Moreover, $\mathcal{S}$ uniquely determines the gaps. For example, suppose J and K are adjacent intervals in $\mathcal{S}$, with the elements of J smaller than these of K. If j and k are the right- and left-hand endpoints of J and K, respectively, then $j < k - 1$ (by the maximality of J), and (j, k) is a non-empty gap in M that is included in I. Similarly, if J is the smallest (or largest) interval in $\mathcal{S}$, and its left-hand (respectively, right-hand) endpoint j differs from the left-hand (respectively, right-hand) endpoint i of I (which is assumed to be in M), then (i, j) (respectively, (j, i)) is a gap in M included in I. This proves (i).

Turning to the proof of (ii), suppose that f is an isomorphism from $\mathfrak{M}$ to $\mathfrak{M}'$. If (k, ℓ) is a gap in $\mathfrak{M}$, then certainly $f(k)$ and $f(\ell)$ are in M' (since $k, \ell \in M$). If there were an integer m strictly between $f(k)$ and $f(\ell)$ in M', then $f^{-1}(m)$ would be strictly between k and ℓ in M, since f^{-1} preserves order. But that is impossible. Hence $(f(k), f(\ell))$ is a gap in M'. This proves (α).

Assume now that I is an interval of Z included in M. If k and ℓ are adjacent in I, say $\ell = k + 1$, then $f(\ell) = f(k) + 1$, since f preserves 1 and the ternary relation $+$. Thus, $f[I]$ is certainly an interval. Suppose, e.g., that $-\infty$ is the left-hand endpoint of I. Then $I \cap (-\infty, -1]$ is the union of a strictly increasing infinite chain of finite intervals. Because f preserves $-\infty, -1, <$, and the property of being an interval, we see that $f[I] \cap (-\infty, -1]$ also the union of such a chain. Thus, $-\infty$ is the left-hand endpoint of $f[I]$. Other types of interval are handled similarly. Thus, $f[I]$ is the same type of interval as I. If $f[I]$ is properly included in an interval $J \subseteq M'$, then $f^{-1}[J]$ is an interval in M that properly includes I. Hence, if I is maximal in M, then $f[I]$ is maximal in M'. This completes the proof of (β).

Part (γ) is an immediate consequence of (i) and (ii)(α),(β). □

Taking $I = (-\infty, \infty)$ in part (i), we see that $M \sim \{-\infty, \infty\}$ itself can be thought of as a series of maximal intervals and gaps in M. Because $\mathfrak{M}$ is closed under $-$, these intervals will be symmetric about 0. By the previous lemma, an isomorphism f between $\mathfrak{M}$ and another structure $\mathfrak{M}'$ of $\mathfrak{Z}_e$ must preserve the basic types of these intervals, including their length. In particular, the interval containing -1, 0, and 1 is mapped identically to itself (since f preserves -1, 0, 1, and $+$). However, f has a certain degree of freedom to expand or contract gaps. (See Figure 4.6.)

DEFINITION 4.6. Let $\mathfrak{M}$ and $\mathfrak{M}'$ be substructures of $\mathfrak{Z}_e$, $\tau(v_0, \ldots, v_{n-1})$ a relation algebraic term, and $a, a' \in {}^nA$.

(i) $\mathfrak{M}$ *contains the indicators of* $\tau[a]$ if $In_{\tau[a]} \subseteq M$.

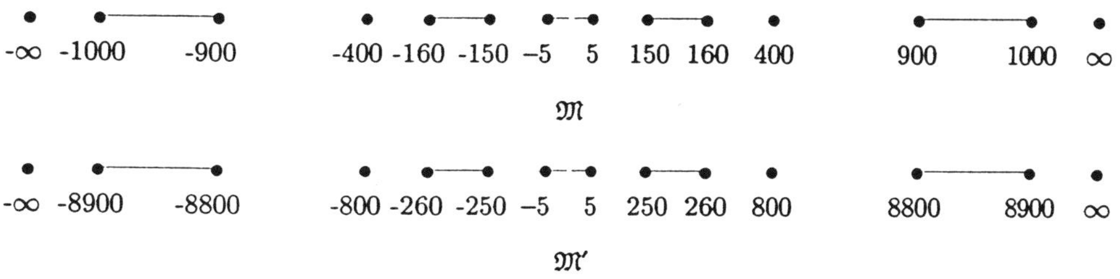

FIGURE 4.6. Isomorphic structures $\mathfrak{M}$ and $\mathfrak{M}'$

(ii) Suppose that $\mathfrak{M}$ and $\mathfrak{M}'$ contain the indicators of $\tau[a]$ and $\tau[a']$, respectively. An isomorphism f from $\mathfrak{M}$ to $\mathfrak{M}'$ *preserves* the indicators of $(\tau[a], \tau[a'])$ if

$$f(h(\tau[a])) = h(\tau[a'])$$

for each of the indicator functions $h = P, p, N, n, z$. □

The connection between the problem of satisfaction of an inequality $\tau \neq 0$ in $\mathfrak{A}$ and the substructures of $\mathfrak{Z}_e$ is given by the following lemma.

LEMMA 4.7. *Let $a, a' \in {}^nA$, let $\tau(v_0, \ldots, v_{n-1})$ be a relation algebraic term, and let $\mathfrak{M}$, $\mathfrak{M}'$ be substructures of $\mathfrak{Z}_e$ that contain the indicators of $\tau[a]$ and $\tau[a']$ respectively. If there is an isomorphism between $\mathfrak{M}$ and $\mathfrak{M}'$ preserving the indicators of $(\tau[a], \tau[a'])$, then $\tau[a] \neq \varnothing$ iff $\tau[a'] \neq \varnothing$.*

PROOF. Notice that

(1) $\quad \tau[a] \neq \varnothing$ iff at least one of its indicators is non-zero.

Indeed, if $\tau[a] = \varnothing$, then all of its indicators are 0, by Definition 4.3. If $\tau[a] \neq \varnothing$, then there is some m in $\tau[a]$. If $m = 0$, then $z(\tau[a]) = 1$. If $m < 0$, then both $N(\tau[a])$ and $n(\tau[a])$ are non-zero. If $m > 0$, then both $P(\tau[a])$ and $p(\tau[a])$ are non-zero. This proves (1).

In view of (1), we also have

(2) $\quad \tau[a'] \neq \varnothing$ iff at least one of its indicators is non-zero.

Suppose there is an isomorphism between $\mathfrak{M}$ and $\mathfrak{M}'$ preserving the indicators of $(\tau[a], \tau[a'])$. The isomorphism also preserves 0, so one of the indicators of $\tau[a]$ will be non-zero iff the corresponding indicator of $\tau[a']$ is non-zero. □

The previous lemma allows us to translate questions about satisfaction in $\mathfrak{A}$ to questions about preservation of indicators by isomorphisms between substructures of $\mathfrak{Z}_e$. Clearly, one important task is to give fairly simple conditions which imply that an isomorphism preserves the indicators of $(\tau[a], \tau[a'])$. This is done in Lemma 4.10 below. The following definitions and Lemma are preparatory.

Let $a \in {}^nA$. By a *(primitive) constituent* of a we mean an intersection of the form $\bigcap_{i<n} b_i$, where, for each $i < n$, either $b_i = a_i$ or $b_i = {\sim}a_i$. In case $b_i = a_i$, we speak of a *positive occurrence* of a_i in the constituent; in case $b_i = {\sim}a_i$, we say that the occurrence is *negative*. As is well known, a_i is the union of all those constituents of a in which it occurs positively, and ${\sim}a_i$ is the union of all those constituents in which a_i occurs negatively. Thus, the union of all the constituents of a is 1. Two different constituents of a have, for some $i < n$, occurrences of a_i of opposite sign; thus, the intersection of two different constituents of a is empty. To give an example, in the case $n = 2$, a has 4 pairwise disjoint constituents that sum to 1:

$$a_0 \cap a_1 \quad , \quad a_0 \cap {\sim}a_1 \quad , \quad {\sim}a_0 \cap a_1 \quad , \quad {\sim}a_0 \cap {\sim}a_1 \, .$$

The sum of the first two, in which a_0 occurs positively, is a_0. The sum of the last two, in which a_0 occurs negatively, is ${\sim}a_0$.

If $a, a' \in {}^nA$, then a constituent $\bigcap_{i<n} b_i$ of a and a constituent $\bigcap_{i<n} c_i$ of a' *correspond* iff, for each $i < n$, we have $b_i = a_i$ iff $c_i = a'_i$ (and, thus, $b_i = {\sim}a_i$ iff $c_i = {\sim}a'_i$). Finally, we use the symbolism $a^{\smile}$ to denote the sequence $\langle a_i^{\smile} : i < n \rangle$.

LEMMA 4.8. *Let f be an isomorphism between two substructures $\mathfrak{M}$, $\mathfrak{M}'$ of $\mathfrak{Z}_e$. Suppose that $a, a' \in {}^nA$ satisfy the following conditions*:

(P1) *For any constituent of a, any constituent of $a^{\smile}$, and any gap (k, ℓ) of $\mathfrak{M}$, the intersection of the two constituents and the gap (k, ℓ) is empty iff the intersection of the corresponding constituents of a' and $a'^{\smile}$ and the gap $(f(k), f(\ell))$ in $\mathfrak{M}'$ is empty*;

(P2) *For each $m \in M$ and $i < n$, we have $m \in a_i$ iff $f(m) \in a'_i$.*

Then for every interval I of Z with endpoints k, ℓ in M, and for its corresponding interval I' under f (of the same type as I, with endpoints $f(k)$ and $f(\ell)$), we have

$$I \cap a_i = \varnothing \quad \textit{iff} \quad I' \cap a'_i = \varnothing \quad , \quad \textit{and} \quad I \cap a_i^{\smile} = \varnothing \quad \textit{iff} \quad I' \cap a'^{\smile}_i = \varnothing \, .$$

PROOF. We begin by showing that the conclusion of the lemma holds when I is a gap in $\mathfrak{M}$. Assume that $I \cap a_i = \varnothing$. Then the intersection of any constituent of $a^{\smile}$, any constituent of a in which a_i occurs positively, and I must be empty. Hence, the intersection of the corresponding constituent of $a'^{\smile}$, the corresponding constituent of a', and I' must be empty, by assumption (P1). Summing over all constituents of $a'^{\smile}$ and over all constituents of a' in which a'_i occurs positively, we conclude that $I' \cap a'_i = \varnothing$. Interchanging I, a, and $a^{\smile}$ with I', a', and $a'^{\smile}$ in this argument, we get that $I' \cap a'_i = \varnothing$ implies $I \cap a_i = \varnothing$. In a similar fashion, we see that $I \cap a_i^{\smile} = \varnothing$ iff $I' \cap a_i^{\smile} = \varnothing$.

Next, we show that the conclusion of the lemma holds when I is an interval included in M. In this case I' is an interval included in M', by Lemma 4.5(ii)(β). If $I \cap a_i \neq \varnothing$ — say m is in $I \cap a_i$ — then m must be in M, since $I \subseteq M$. Hence, using property (P2) we see that $f(m)$ is in $I' \cap a'_i$. Thus, $I' \cap a'_i \neq \varnothing$. Again, the reverse implication, and the equivalence $I \cap a_i^{\smile} \neq \varnothing$ iff $I' \cap a'^{\smile}_i \neq \varnothing$, follow readily by similar arguments.

Finally, let I be an arbitrary interval with endpoints in $\mathfrak{M}$. By Lemma 4.5(i), there is a decomposition S of I into the disjoint union of gaps in M and maximal subintervals in $M \cap I$. By 4.5(ii)(γ), f induces a mapping of $\mathcal{S}$ to the decomposition $\mathcal{S}'$ of I' over M'. We have seen that the conclusion of our theorem holds for each gap and interval in M that is contained in $\mathcal{S}$, and the corresponding gap and interval in M' contained in $\mathcal{S}'$. Hence, it holds for $\bigcup \mathcal{S}$ and $\bigcup \mathcal{S}'$, i.e., it holds for I and I'. □

DEFINITION 4.9. Fix $\mathfrak{M}$ and $\mathfrak{M}'$, isomorphic substructures of $\mathfrak{Z}_e$ under a mapping f. Let $a, a' \in {}^nA$. We say that $m, m' \in Z_e$ are *similar* (with respect to a and a') if they satisfy the following conditions:

(i) For each $i < n$, we have $m \in a_i$ iff $m' \in {a'}_i$;
(ii) For each $i < n$, we have $m \in a_i^{\smile}$ iff $m' \in {a'}_i^{\smile}$;
(iii) If $m \in M$, then $f(m) = m'$;
(iv) If (k, ℓ) is a gap in M, then $m \in (k, \ell)$ iff $m' \in (f(k), f(\ell))$. □

REMARKS. At most one of the two conditions (iii) and (iv) in 4.9 will be applicable to any given pair (m, m'). Indeed, each integer m is either in M — in which case condition (iii) is applicable — or in a unique gap in M — in which case condition (iv) is applicable.

Notice that condition (P1) in Lemma 4.8 just says: for every m in a gap in M, there will be an m' in the corresponding gap in M' (under f) that is similar to m, and conversely. Condition (P2) says: each m in M is similar to $f(m)$. Thus, under the hypotheses (P1) and (P2) we conclude that for every integer m there is an integer m' that is similar to m; either $m' = f(m)$ (which happens just in case m is in M) or m and m' are in corresponding gaps.

Finally, observe that if m and m' are similar, then so are $-m$ and $-m'$. For example, to verify condition 4.9(i) for $-m$ and $-m'$ we use the definition of $^{\smile}$ in $\mathfrak{A}$ and 4.9(ii) for m and m':

$$\begin{aligned} -m \in a_i \quad &\text{iff} \quad m \in a_i^{\smile} \\ &\text{iff} \quad m' \in {a'_i}^{\smile} \\ &\text{iff} \quad -m' \in a'_i \,. \end{aligned}$$

To verify condition 4.9(iv) for $-m$ and $-m'$, observe that (k, ℓ) is a gap in M iff $(-\ell, -k)$ is a gap in M; this is because M is closed under negation. Therefore, using condition 4.9(iv) for m and m', and also the fact that f preserves negation, we see that

$$\begin{aligned} -m \in (k, \ell) \quad &\text{iff} \quad m \in (-\ell, -k) \\ &\text{iff} \quad m' \in (f(-\ell), f(-k)) \\ &\text{iff} \quad m' \in (-f(\ell), -f(k)) \\ &\text{iff} \quad -m' \in (f(k), f(\ell)) \,. \qquad \square \end{aligned}$$

LEMMA 4.10. *Let f be an isomorphism between two finite substructures, $\mathfrak{M}$ and $\mathfrak{M}'$, of $\mathfrak{Z}_e$. Suppose that $a, a' \in {}^nA$ satisfy conditions* (P1) *and* (P2) *from* 4.8 *and the following condition*:

(P3) *For every $i < n$,*

$$N(a_i) = -\infty \quad \textit{iff} \quad N(a'_i) = -\infty \quad , \quad \textit{and} \quad P(a_i) = \infty \quad \textit{iff} \quad P(a'_i) = \infty .$$

Let $\tau(v_0, \ldots, v_{n-1})$ be a relation algebraic term, and suppose that $\mathfrak{M}$ contains the indicators of $\sigma[a]$ for every subterm σ of τ. Then, for each subterm σ of τ and for each indicator function h,

(i) $$f(h(\sigma[a])) = h(\sigma[a']) ,$$

i.e., f preserves the indicators of $(\sigma[a], \sigma[a'])$; in particular, $\mathfrak{M}'$ must contain the indicators of $\sigma[a']$.

PROOF. Assume the hypothesis of the lemma. In addition to establishing (i), we shall simultaneously prove:

(1) Whenever $m, m' \in Z$ are similar, we have $m \in \sigma[a]$ iff $m' \in \sigma[a']$.

Notice that

(2) If $m \in M$, then m and $f(m)$ are similar.

Indeed,

$$m \in a_i \quad \text{iff} \quad f(m) \in a_i ,$$

by condition (P2). Thus, condition 4.9(i) is satisfied. Also,

$$\begin{aligned} m \in a_i^{\smile} \quad & \text{iff} \quad -m \in a_i , \\ & \text{iff} \quad f(-m) \in a'_i , \\ & \text{iff} \quad -f(m) \in a'_i , \\ & \text{iff} \quad f(m) \in a_i'^{\smile} , \end{aligned}$$

by definition of $a_i^{\smile}$ and $a'^{\smile}_i$, and by the fact that f preserves $-$. Therefore, 4.9(ii) is satisfied. Condition 4.9(iii) holds trivially, and 4.9(iv) holds vacuously, since m is in M.

The proof of (i) and of (1) proceeds by induction on terms.

BASE CASE: σ is a variable v_i.

Then $\sigma[a] = a_i$ and $\sigma[a'] = a'_i$. In this case, (1) follows directly from condition 4.9(i). To establish (i), we must consider separately the case of each indicator function. We begin with N. Either $N(a_i) = -\infty$, or $N(a_i) = 0$, or $N(a_i)$ is the smallest negative integer in a_i. In the first case, $N(a'_i) = -\infty$ by condition (P3); and certainly f preserves $-\infty$, since ∞ is a distinguished element of $\mathfrak{M}$. In the second case, a_i has no negative numbers, by definition of N, i.e., $(-\infty, 0) \cap a_i = \varnothing$. By Lemma 4.8, $(-\infty, 0) \cap a'_i = \varnothing$. Therefore, $N(a'_i) = 0$, and of course f preserves 0. Finally, suppose that $N(a_i)$ is the smallest negative number in a_i. Setting $k = N(a_i)$, we see that $(-\infty, k) \cap a_i = \varnothing$. Also, $k \in M$, since M contains the indicators of $\sigma[a]$ by assumption. Therefore, $(-\infty, f(k)) \cap a'_i = \varnothing$, by Lemma 4.8. Moreover, $f(k)$ is in a_i', by (P2), because we have assumed that k is in a_i. Further, $k < 0$, so $f(k) < 0$ (since f preserves 0 and $<$). We conclude that $f(k)$ is the smallest negative integer in a'_i. But then $f(k) = N(a'_i)$, by definition of N.

The argument for the case of the indicator function n is similar, so we only present it in outline. Suppose, first, that $n(a_i) = 0$. Then $N(a_i) = 0$, by the remarks preceding the proof of Lemma 4.4. Therefore, $N(a_i') = 0$ by the results of the preceding paragraph. It follows that $n(a_i') = 0$, and f certainly preserves 0. Suppose, next, that $n(a_i) \neq 0$. Then $n(a_i)$ is the largest negative number in a_i. Setting $k = n(a_i)$, we have $k \in M$ and $(k, 0) \cap a_i = \varnothing$. Using Lemma 4.8, condition (P2), and the fact that f preserves $<$ and 0, we see that

$$(f(k), 0) \cap a'_i = \varnothing \quad , \quad f(k) \in a'_i \quad , \quad \text{and} \quad f(k) < 0 .$$

Thus, $f(k)$ is the largest negative number in a'_i. By definition of n, we conclude that $f(k) = n(a'_i)$.

The arguments for the indicators P and p are completely analogous, and the argument for z is quite simple. We leave them all to the reader.

BASE CASE: σ is 1'.

This step is trivial: all the indicators of $\{0\}$ are 0, except that $z(\{0\}) = 1$, and f preserves 0 and 1.

INDUCTIVE STEP: σ is $\eta^{\smile}$.

To establish (1) in this case, assume that m and m' are similar. By the induction hypothesis (1) for $\eta[a]$, but applied to $-m$ and $-m'$, we have

$$-m \in \eta[a] \quad \text{iff} \quad -m' \in \eta[a']$$

(see the remarks following Definition 4.9). Using the definition of $^{\smile}$ in $\mathfrak{A}$ and the definition of $\tau[a]$, we also have

$$\begin{aligned} \eta^{\smile}[a] &= \eta[a]^{\smile} = \{k : -k \in \eta[a]\} \\ \eta^{\smile}[a'] &= \eta[a']^{\smile} = \{k : -k \in \eta[a']\} . \end{aligned}$$

Combining these observations, we get (1):

$$m \in \eta^{\smile}[a] \quad \text{iff} \quad m' \in \eta^{\smile}[a'] .$$

For (i) we treat only the case of the indicator function N. Observe that

$$N(\eta^{\smile}[a]) = N(\eta[a]^{\smile}) = -P(\eta[a]) \quad \text{and} \quad N(\eta^{\smile}[a']) = N(\eta[a']^{\smile}) = -P(\eta[a']) .$$

Therefore, using the induction hypothesis (i) for $\eta[a]$ and the preservation properties of f we arrive at

$$f(N(\eta^{\smile}[a])) = f(-P(\eta[a])) = -f(P(\eta[a])) = -P(\eta[a']) = N(\eta^{\smile}[a']) .$$

INDUCTIVE STEP: σ is $-\eta$.

This is the most difficult inductive step of the proof. Suppose that σ is $-\eta$. Our induction hypothesis is that (i) and (1) hold for η (in place of σ). To establish (1), assume that m and m' are similar. Then

$$\begin{aligned} m \in -\eta[a] \quad &\text{iff} \quad m \notin \eta[a] && \text{by definition of } -\eta[a], \\ &\text{iff} \quad m' \notin \eta[a'] && \text{by the induction hypothesis (1) for } \eta, \\ &\text{iff} \quad m' \in -\eta[a'] && \text{by definition of } -\eta[a']. \end{aligned}$$

We turn to the verification of (i). By the induction hypothesis (i) for η,

(3) $h(\eta[a']) = f(h(\eta[a]))$ for every indicator function h.

In particular, since f preserves $-\infty$, ∞, -1, 1, 0, and $<$, we have

(4) For h an indicator function, $h(\eta[a])$ assumes one of the values $-\infty$, ∞, -1, 1, 0 iff $h(\eta[a'])$ assumes the same value; $h(\eta[a])$ is in $(-\infty, -1)$ iff $h(\eta[a'])$ is in $(-\infty, -1)$; and $h(\eta[a])$ is in $(1, \infty)$ iff $h(\eta[a'])$ is in $(1, \infty)$.

We treat in detail the case of the indicator function N. Our goal is to prove that

(5) $f(N(-\eta[a])) = N(-\eta[a'])$.

CASE 1: $N(\eta[a]) = 0$.

In this case, $\eta[a]$ has no negative numbers, i.e., $(-\infty, 0) \subseteq -\eta[a]$. By (4), $N(\eta[a']) = 0$. Therefore, $(-\infty, 0) \subseteq -\eta[a']$. Thus,

$$N(-\eta[a]) = -\infty = N(-\eta[a']), \tag{6}$$

so (5) must hold.

CASE 2: $N(\eta[a])$ is in $(-\infty, -1]$.

Here, $N(\eta[a])$ is the smallest negative number in $\eta[a]$. By (4), $N(\eta[a'])$ is in $(-\infty, -1]$, and is therefore the smallest negative number in $\eta[a']$. Consequently,

$$(-\infty, N(\eta[a])) \subseteq -\eta[a] \quad \text{and} \quad (-\infty, N(\eta[a'])) \subseteq -\eta[a'].$$

Again, we conclude that (6), and hence (5), holds.

CASE 3: $N(\eta[a]) = -\infty$.

This is the main case. Notice that, in this case, $\eta[a]$ must contain arbitrarily small integers. Since $\eta[a]$ is a finite or cofinite subset of Z (it is in A), we conclude that

(7) $\eta[a]$ is cofinite.

Thus, $-\eta[a]$ is finite, and $N(-\eta[a])$ is either 0 (which happens just in case $(-\infty, 0)$ is a subset of $\eta[a]$) or it is the smallest negative integer in $-\eta[a]$, i.e., the smallest integer ≤ -1 that is not in $\eta[a]$. In either case we have

$$(8) \qquad (-\infty, N(-\eta[a])) \subseteq \eta[a]\,.$$

Set $j = f(N(-\eta[a]))$. Our immediate goal is to show that

$$(9) \qquad (-\infty, j) \subseteq \eta[a']\,.$$

Let m' be any integer in $(-\infty, j)$. If $m' \in M'$, then set $m = f^{-1}(m')$. Since f^{-1} preserves $-\infty$ and $<$, and maps j to $N(-\eta[a])$, we see that m is in $(-\infty, N(-\eta[a]))$. Therefore, m is in $\eta[a]$, by (8). But m is similar to m', by (2). Applying the induction hypothesis (1), we conclude that m' is in $\eta[a']$, as desired.

Now suppose that $m' \notin M'$. Then m' must be in some gap (k', ℓ') of M' with $\ell' \leq j$ (since $j \in M'$). Set $k = f^{-1}(k')$ and $\ell = f^{-1}(\ell')$. Then (k, ℓ) is a gap of M, with

$$\ell \leq f^{-1}(j) = N(-\eta[a])\,,$$

by 4.5(ii)(β) applied to f^{-1}, and by the fact that f^{-1} preserves $\leq$. Define $\hat{b}$ and $\hat{c}$ in ${}^{n}A$ by setting

$$\hat{b}_i = \begin{cases} a'_i & \text{if } m' \in a'_i \\ -a'_i & \text{if } m' \notin a'_i \end{cases},$$

$$\hat{c}_i = \begin{cases} a'^{\smile}_i & \text{if } m' \in a'^{\smile}_i \\ -a'^{\smile}_i & \text{if } m' \notin a'^{\smile}_i \end{cases},$$

for each $i < n$. Let b' and c' be the resulting constituents of a' and $a'^{\smile}$:

$$b' = \textstyle\bigcap_{i<n} \hat{b}_i \quad \text{and} \quad c' = \bigcap_{i<n} \hat{c}_i\,,$$

and take b and c to be the corresponding constituents of a and $a^{\smile}$. Then m' is in $b' \cap c' \cap (k', \ell')$ by construction. Therefore, by condition (P1), the set $b \cap c \cap (k, \ell)$ is not empty. Take m to be some element in it. Then m and m' are similar: by their construction, they satisfy conditions 4.9(i),(ii),(iv), and condition 4.9(iii) holds vacuously ($m \notin M$, because it is in the gap (k, ℓ)). Since $m < \ell \leq N(-\eta[a])$, we see that m is in $(-\infty, N(-\eta[a]))$. By (8), m must be in $\eta[a]$. Using the fact that m and m' are similar, and invoking the induction hypothesis (1), we conclude that m' is in $\eta[a']$. This proves (9).

Suppose, now, that $N(-\eta[a]) = 0$, i.e., $-\eta[a]$ contains no negative numbers. By (9), the definition of j, and the fact that f preserves 0, we get $(-\infty, 0) \subseteq \eta[a']$. Hence, $N(-\eta[a']) = 0$, by definition of N. Thus, (5) holds in this case.

Suppose that $N(-\eta[a]) \leq -1$. Then $j \leq -1$, because f preserves $\leq$ and -1. Now $N(-\eta[a])$ and j are similar, by (2) and the definition of j. Also, in this case $N(-\eta[a])$ is not in $\eta[a]$ (see the remark preceding (8)). Therefore, by the induction hypothesis (1), we obtain that j is not in $\eta[a']$. In view of (9), we see that j is the smallest negative integer that is in $-\eta[a']$, i.e., that is not in $\eta[a']$. But this is just the definition of $N(-\eta[a'])$; thus, $N(-\eta[a']) = j$. In view of the definition of j, this completes the proof of (5).

Since the argument in case of the indicator function n is rather similar to that of N, we content ourselves with an outline of the proof that

(10) $\quad f(n(-\eta[a])) = n(-\eta[a'])\,.$

If $n(\eta[a]) = 0$, then $n(\eta[a']) = 0$, by (4). Hence,

$$(-\infty, 0) \subseteq -\eta[a] \quad \text{and} \quad (-\infty, 0) \subseteq -\eta[a']\,.$$

In this case,

(11) $\quad n(-\eta[a]) = -1 = n(-\eta[a'])\,,$

so (10) holds. Similarly, if $n(\eta[a]) < -1$, then $n(\eta[a']) < -1$, by (4). Again, we obtain (11), and hence (10).

Suppose, now, that $n(\eta[a])$, and hence also $n(\eta[a'])$, assumes the value -1. Then $n(-\eta[a])$ is either 0 (which happen just in case $(-\infty, 0) \subseteq \eta[a]$) or it is the largest negative integer in $-\eta[a]$, i.e., the largest integer ≤ -1 that is not in $\eta[a]$. Set $i = -\infty$ in the first case and $i = n(-\eta[a])$ in the second case, and put $j = f(i)$. We have

(12) $\quad (i, 0) \subseteq \eta[a]\,,$

and we want to prove that

(13) $\quad (j, 0) \subseteq \eta[a']\,.$

Let m' be an arbitrary element of $(j, 0)$. If $m' \in M'$, set $m = f^{-1}(m')$. Then m and m' are similar, by (2), and $m \in (i, 0)$. By (12), we have $m \in \eta[a]$, so by the induction hypothesis (1), we obtain $m' \in \eta[a']$.

Now suppose that $m' \notin M'$. Then m' is in some gap (k', ℓ') of M' with

$$j \leq k' < \ell' \leq -1\,.$$

Set $k = f^{-1}(k')$ and $\ell = f^{-1}(\ell')$, and notice that (k, ℓ) is a gap in M with

$$i \leq k < \ell \leq -1\,.$$

Construct constituents b', c', b, and c of a', $a'^{\smile}$, a, and $a^{\smile}$ as before, and observe that m' is in $b' \cap c' \cap (k', \ell')$ by construction. Thus, by condition (P1), the set $b \cap c \cap (k, \ell)$ is not empty. Choose any m in this latter set. Then m is similar to m', by 4.9, and $m \in (i, 0)$. By (12), m is in $\eta[a]$. Therefore, by the induction hypothesis (1), m' is in $\eta[a']$. This proves (13).

If $i = -\infty$, i.e., if $(-\infty, 0) \subseteq \eta[a]$, then $j = -\infty$ and $(-\infty, 0) \subseteq \eta[a']$, by (13). In this case, both $n(-\eta[a])$ and $n(-\eta[a'])$ are 0, so (10) holds. If $-\infty < i \leq -1$, then $-\infty < j \leq -1$, since f preserves $<$. In this case, $i = n(-\eta[a])$, so i is in M and not in $\eta[a]$. Since i and j are similar, by (2), j is not in $\eta[a']$, by the induction hypothesis (1). Together with (13), this shows that j is the largest negative integer not in $\eta[a']$, i.e., $j = n(-\eta[a'])$. In view of the definition of j, this proves (10).

The arguments for the indicator functions P and p are completely analogous to those for N and n, and are left to the reader. The argument for z is quite simple. Notice, that 0 is similar to itself, e.g., by (2). Therefore, by the induction hypothesis (1), we have $0 \in \eta[a]$ iff $0 \in \eta[a']$. Consequently,

$$\begin{aligned} z(-\eta[a]) = 1 \quad &\text{iff} \quad 0 \in -\eta[a] \\ &\text{iff} \quad 0 \notin \eta[a] \\ &\text{iff} \quad 0 \notin \eta[a'] \\ &\text{iff} \quad 0 \in -\eta[a'] \\ &\text{iff} \quad z(-\eta[a']) = 1\,. \end{aligned}$$

INDUCTIVE STEP: σ is $\xi + \eta$.

Assume, as our induction hypothesis, that (i) and (1) hold for ξ and η. Of course,

(14) $\quad (\xi + \eta)[a] = \xi[a] \cup \eta[a]\,.$

Therefore the verification of (1) is easy. Let m and m' be similar. Then

$$\begin{aligned} m \in (\xi+\eta)[a] \quad &\text{iff} \quad m \in \xi[a] \text{ or } m \in \eta[a] && \text{by (14),} \\ &\text{iff} \quad m' \in \xi[a'] \text{ or } m' \in \eta[a'] && \text{by the induction hypothesis (1),} \\ &\text{iff} \quad m' \in (\xi+\eta)[a'] && \text{by the analogue of (14) for } a'\,. \end{aligned}$$

We verify (i) in the case of the indicator function N. The other cases are equally easy, and are left to the reader. From (14) and the definition of N, we see that

$$N(\xi + \eta[a]) = \min\{N(\xi[a]), N(\eta[a])\}\,.$$

Without loss of generality we may assume that

$$N(\xi[a]) \leq N(\eta[a])\,.$$

Then $N(\xi + \eta[a]) = N(\xi[a])$. Since f preserves $\leq$, we obtain

(15) $\quad f(N(\xi + \eta[a])) = f(N(\xi[a])) \leq f(N(\eta[a]))\,.$

Applying the induction hypothesis (i) to (15), we have

(16) $\quad N(\xi[a']) = f(N(\xi[a])) \leq f(N(\eta[a])) = N(\eta[a'])\,.$

Therefore,

$$N(\xi + \eta[a']) = \min\{N(\xi[a']), N(\eta[a'])\} = N(\xi[a'])\,.$$

In view of (15) and (16), we conclude that

$$N(\xi + \eta[a']) = f(N(\xi + \eta[a]))\,.$$

INDUCTIVE STEP: σ is $\xi\,;\eta$.

Assume that (i) and (1) hold for ξ and η in place of σ. In principle, we must verify (1) and (i) for σ in each of the cases discussed in the arithmetic rules of Lemma 4.4 and also in each of the cases where one of $\xi[a]$ and $\eta[a]$ coincides with 0 or 1'. However — and this is the whole point — f *must* preserve the indicators of $(\xi\,;\eta[a]\,,\,\xi\,;\eta[a'])$: they are arithmetically computable from the indicators of $(\xi[a],\xi[a'])$ and $(\eta[a],\eta[a'])$, and f is assumed to preserve these latter indicators. Thus, it should suffice to work out just a few cases as examples.

To begin with, assume that $\xi[a] = 1'$. Then

$$n(\xi[a]) = p(\xi[a]) = 0 \quad \text{and} \quad z(\xi[a]) = 1\,.$$

By the induction hypothesis (i) and the fact that f preserves 0 and 1, we have

$$n(\xi[a']) = p(\xi[a']) = 0 \quad \text{and} \quad z(\xi[a']) = 1\,,$$

i.e., $\xi[a'] = 1'$. Hence,

$$\eta[a] = \xi[a]\,;\eta[a] = \xi\,;\eta[a] \quad \text{and} \quad \eta[a'] = \xi[a']\,;\eta[a'] = \xi\,;\eta[a']\,.$$

Thus, the indicators of $\xi\,;\eta[a]$ and $\xi\,;\eta[a']$ coincide with those of $\eta[a]$ and $\eta[a']$ respectively. Since f preserves the indicators of the latter, by the induction hypothesis (i), it must preserve the indicators of the former. Thus, (i) holds. In a similar fashion, we see that (1) also holds. The other trivial cases when one of $\xi[a]$ and $\eta[a]$ is 0 or 1' are handled analogously.

We now assume that both $\xi[a]$ and $\eta[a]$ are different from 0 and from 1'.

Suppose that

$$n(\xi[a]) = n(\eta[a]) = 0 \qquad \text{and} \quad z(\xi[a]) = z(\eta[a]) = 1$$

(see 4.4(v)). Then

$$n(\xi[a']) = n(\eta[a']) = 0 \quad \text{and} \quad z(\xi[a']) = z(\eta[a']) = 1\,,$$

by the induction hypothesis (i). Further,

$$P(\xi[a]) \quad , \quad P(\eta[a]) \quad , \quad P(\xi[a']) \quad , \quad P(\eta[a'])$$

are all positive (see the remark preceding the proof of Lemma 4.4). Therefore, by 4.4(v), we obtain

(17) $(\xi\,;\eta)[a] = [0, P(\xi[a]) + P(\eta[a]) + 1)\,,$

(18) $(\xi\,;\eta)[a'] = [0, P(\xi[a']) + P(\eta[a']) + 1)\,.$

Now using Definition 4.3, we immediately compute that

$$\begin{aligned} z(\xi\,;\eta[a]) &= z(\xi\,;\eta[a']) &&= 1\,,\\ p(\xi\,;\eta[a]) &= p(\xi\,;\eta[a']) &&= 1\,,\\ n(\xi\,;\eta[a]) &= n(\xi\,;\eta[a']) &&= 0\,,\\ N(\xi\,;\eta[a]) &= N(\xi\,;\eta[a']) &&= 0\,. \end{aligned}$$

Since f preserves 0 and 1, it obviously preserves z, p, n, and N in this case. From 4.3 and (17), (18) we have

(19) $P(\xi\,;\eta[a]) = P(\xi[a]) + P(\eta[a])\,,$

(20) $P(\xi\,;\eta[a']) = P(\xi[a']) + P(\eta[a'])\,,$

Because

$$f(P(\xi[a])) = P(\xi[a']) \quad \text{and} \quad f(P(\eta[a])) = P(\eta[a'])\,,$$

by the induction hypothesis (i), and because f preserves $+$, we see from (19), (20), and the assumption that all indicators are in $\mathfrak{M}$ that

(21) $f(P(\xi\,;\eta[a])) = P(\xi\,;\eta[a'])\,.$

Thus, (i) holds for $\xi\,;\eta$ in this case.

To check (1), suppose that m and m' are similar. If $m \in M$, then $m' = f(m)$, by 4.9(iii). Therefore,

$m \in \xi\,;\eta[a]$	iff	$0 \le m \le P(\xi\,;\eta[a])$	by (17), (19),
	iff	$0 \le m' \le P(\xi\,;\eta[a'])$	by (21) and the preservation properties of f,
	iff	$m' \in \xi\,;\eta[a']$	by (18), (20).

If $m \notin M$, then m is in some gap (k,ℓ) of M, and m' is in the corresponding gap $(f(k), f(\ell))$ of M', by 4.9(iv). Since 0 and $P(\xi\,;\eta[a])$ are in M, it follows from the definition of gap that

$m \in \xi\,;\eta[a]$	iff	$0 \le m \le P(\xi\,;\eta[a])$	by (17), (19),
	iff	$(k,\ell) \subseteq (0, P(\xi\,;\eta[a]))$	by the definition of gap and the fact that 0 and $P(\xi\,;\eta[a])$ are in M,
	iff	$(f(k), f(\ell)) \subseteq (0, P(\xi\,;\eta[a']))$	by (21) and the preservation properties of f,
	iff	$0 \le m' \le P(\xi\,;\eta[a'])$	by the definition of gap and the fact that 0 and $P(\xi\,;\eta[a'])$ are in M',
	iff	$m' \in \xi\,;\eta[a']$	by (18), (20).

This completes the verification of (1) in the case under consideration.

For the next case, suppose that

$$p(\xi[a]) = 0 \quad , \quad n(\eta[a]) = 0 \quad , \quad \text{and} \quad n(\xi[a]) + P(\eta[a]) < 0\,.$$

(see 4.4(iv)). Then

$$p(\xi[a']) = 0 \quad , \quad n(\eta[a']) = 0 \quad , \quad \text{and} \quad n(\xi[a']) + P(\eta[a']) < 0\,,$$

by the induction hypothesis (i) and the preservation properties of f. Applying 4.4(iv) to $(\xi\,;\eta)[a]$ and $(\xi\,;\eta)[a']$, we get

$$(22)\quad (\xi\,;\eta)[a] = (-\infty, n(\xi[a]) + P(\eta[a])] \cup [1,\infty)\,,$$

$$(23)\quad (\xi\,;\eta)[a'] = (-\infty, n(\xi[a']) + P(\eta[a'])] \cup [1,\infty)\,.$$

From 4.3 we see that N, P, p, and z will assume, respectively, the values $-\infty$, ∞, 1, and 0 on $(\xi\,;\eta)[a]$ and $(\xi\,;\eta)[a']$. Since f preserves each of these values, it will preserve each of the appropriate indicators. From 4.3, (22), (23), and the case that we are in, we also have

$$(24)\quad n(\xi\,;\eta[a]) = n(\xi[a]) + P(\eta[a])\,,$$

$$(25)\quad n(\xi\,;\eta[a']) = n(\xi[a']) + P(\eta[a'])\,.$$

Because

$$f(n(\xi[a])) = n(\xi[a']) \quad \text{and} \quad f(P(\eta[a])) = P(\eta[a'])\,,$$

by the induction hypothesis (i), and because f preserves $+$, we see from (24) and (25) that

$$(26)\quad f(n(\xi\,;\eta[a])) = n(\xi\,;\eta[a'])\,.$$

This shows that (i) holds in the present case.

To prove (1), suppose that m and m' are similar. If $m \in M$, then $f(m) \in M'$, by 4.9(iii). In this case,

$m \in \xi\,;\eta[a]$ iff $m \in (-\infty, n(\xi\,;\eta[a]))$ or $m \in [1,\infty)$	by (22), (24),
iff $m' \in (-\infty, n(\xi\,;\eta[a']))$ or $m' \in [1,\infty)$	by (26) and the preservation properties of f,
iff $m' \in \xi\,;\eta[a']$	by (23), (25).

If $m \notin M$, then there is a gap (k,ℓ) in M that contains m. In this case, $(f(k), f(\ell))$ is a gap in M' that contains m', by condition 4.9(iv). Therefore, using the definition of a gap and the fact that $-\infty$, 1, ∞, and $n(\xi\,;\eta[a])$ are all in M, we get

$m \in \xi\,;\eta[a]$ iff $(k,\ell) \subseteq (-\infty, n(\xi\,;\eta[a]))$ or $(k,\ell) \subseteq [1,\infty)$	by (22), (24),
iff $(f(k), f(\ell)) \subseteq (-\infty, n(\xi\,;\eta[a']))$ or $(f(k), f(\ell)) \subseteq [1,\infty)$	by (26) and the preservation properties of f,
iff $m \in \xi\,;\eta[a']$	by (23), (25).

This completes the verification of (1) in this case.

For our last example, we treat the case when

$$(27)\quad h(\xi[a]) \neq 0 \text{ and } h(\eta[a]) \neq 0 \text{ for each of the indicator functions } h = p, n.$$

(See 4.4(i).) By the induction hypothesis (i) and the fact that f preserves 0, we see that

(28) $h(\xi[a']) \neq 0$ and $h(\eta[a']) \neq 0$ for each of the indicator functions $h = p, n$.

Let m be any integer and let m' be an integer similar to m; such an m' must exist by the remarks following Definition 4.9. Using the induction hypothesis (1) and the induction step from η to $\eta^{\smile}$, we easily get

$$\begin{aligned} m \in \xi[a] \cap \eta[a]^{\smile} \quad &\text{iff} \quad m \in \xi[a] \text{ and } m \in \eta^{\smile}[a] \\ &\text{iff} \quad m' \in \xi[a'] \text{ and } m' \in \eta^{\smile}[a'] \\ &\text{iff} \quad m' \in \xi[a'] \cap \eta[a']^{\smile} . \end{aligned}$$

In other words,

(29) $\xi[a] \cap \eta[a]^{\smile} \neq 0 \quad \text{iff} \quad \xi[a'] \cap \eta[a']^{\smile} \neq 0 .$

Combining (27)–(29), we arrive at

(30) $\xi \,;\eta[a] = \xi \,;\eta[a'] = 0\text{'} \quad \text{or} \quad \xi \,;\eta[a] = \xi \,;\eta[a'] = 1 .$

For example,

$$\begin{aligned} \xi \,;\eta[a] = 1 \quad &\text{iff} \quad \xi[a] \cap \eta[a]^{\smile} \neq 0 &&\text{by (27) and 4.4(i),} \\ &\text{iff} \quad \xi[a'] \cap \eta[a']^{\smile} \neq 0 &&\text{by (29),} \\ &\text{iff} \quad \xi \,;\eta[a'] = 1 &&\text{by (28) and 4.4(i).} \end{aligned}$$

It follows from (30) that $\xi \,;\eta[a]$ and $\xi \,;\eta[a']$ have the same indicators and that these indicators are either distinguished constants of $\mathfrak{M}$ and $\mathfrak{M}'$ or negations of such constants. For instance,

$$N(\xi \,;\eta[a]) = N(\xi \,;\eta[a']) = -\infty .$$

Since f preserves all such constants, it must preserve these indicators. This proves that (i) holds in this case. To verify (1), suppose that m and m' are similar. If $m = 0$, then m is in M and hence $m' = f(m) = 0$, by 4.9(iii) and the preservation properties of f. Similarly, if $m \neq 0$, then $m' \neq 0$. From (30) we at once conclude that

$$m \in \xi \,;\eta[a] \quad \text{iff} \quad m' \in \xi \,;\eta[a'] .$$

This completes the verification of (1) in this case, and hence the proof of the lemma. □

LEMMA 4.11. *Let f be an isomorphism between two finite substructures, $\mathfrak{M}$ and $\mathfrak{M}'$, of $\mathfrak{Z}_e$ that satisfies the following condition with respect to a given natural number n:*

(P4) *If (k, ℓ) is any gap in M with integer endpoints, and if $f(\ell) - f(k) < 2^{2n}$, then $f(\ell) - f(k) = \ell - k$.*

Suppose $a \in {}^nA$ and $\tau(v_0, \dots, v_{n-1})$ is a relation algebraic term such that $\mathfrak{M}$ contains the indicators of $\sigma[a]$ for every subterm σ of τ. Then there is an $a' \in {}^nA$ with the following properties:

(i) *For each $i < n$, the intervals $(-\infty, -s)$ and (s, ∞) are either both included in a'_i or both disjoint from a'_i, where $s = \max(M' \sim \{\infty\}) + 2^{2n}$; the former happens iff a_i is cofinite;*

(ii) *For each subterm σ of τ, the isomorphism f preserves the indicators of $(\sigma[a], \sigma[a'])$.*

In particular, $\mathfrak{M}'$ contains the indicators of $\sigma[a']$ for every subterm σ of τ.

PROOF. We shall construct an $a' \in {}^nA$ that satisfies the conditions (P1)–(P3) given in 4.8 and 4.10. Imitating the definition of similarity in 4.9, we define an equivalence relation on Z. Two integers p, q are equivalent, in symbols $p \simeq q$, iff the following four conditions hold:

(1) For each $i < n$, we have $p \in a_i$ iff $q \in a_i$;

(2) For each $i < n$, we have $p \in a_i^{\smile}$ iff $q \in a_i^{\smile}$;

(3) If one of p, q is in M, then $p = q$;

(4) If (k, ℓ) is a gap in M, then $p \in (k, \ell)$ iff $q \in (k, \ell)$.

We see from (3) that $\simeq$ restricted to $M \sim \{-\infty, \infty\}$ is the identity relation. Together, conditions (1), (2), and (4) amount to saying that, for every constituent b of a, every constituent c of $a^{\smile}$, and every gap (k, ℓ) of M,

$$p \in b \cap c \cap (k, \ell) \quad \text{iff} \quad q \in b \cap c \cap (k, \ell)$$

(see condition (P1) in Lemma 4.8). In other words, the equivalence classes of $\simeq$ are the singletons $\{p\}$, where p ranges over the integers in M, and the non-empty sets $b \cap c \cap (k, \ell)$, where b, c, (k, ℓ) range over the constituents of a, the constituents of $a^{\smile}$, and the gaps in M, respectively. Each of a and $a^{\smile}$ has at most 2^n constituents. Thus, there are at most $2^n \cdot 2^n$, or 2^{2n}, pairs (b, c) of constituents of a and $a^{\smile}$. It follows that each gap in M is the union of at most 2^{2n} equivalence classes of $\simeq$.

In defining a'_i it suffices to concentrate on positive gaps (k, ℓ), i.e., gaps in which $1 \le k$ and $\ell \le \infty$. Indeed, if b and c are constituents of a and $a^{\smile}$, respectively, then the equivalence class $b \cap c \cap (-\ell, -k)$ is just the set of negatives of integers in the equivalence class $[b \cap c \cap (-\ell, -k)]^{\smile}$, which is of course the same as the equivalence class $b^{\smile} \cap c^{\smile} \cap (k, \ell)$; moreover, $b^{\smile}$ and $c^{\smile}$ are constituents of $a^{\smile}$ and a, respectively. In other words, the function $m \mapsto -m$ induces a bijection between the equivalence classes of $\simeq$ that partition (k, ℓ) and the equivalence classes of $\simeq$ that partition $(-\ell, -k)$.

Let (k, ℓ) be a positive gap in M with integer endpoints. If $f(\ell) - f(k) < 2^{2n}$, then take $g_{k,\ell}$ to be any bijection of (k, ℓ) onto $(f(k), f(\ell))$. Such a bijection exists by condition (P4). If $f(\ell) - f(k) \ge 2^{2n}$, then let m be the number of equivalence classes of $\simeq$ in (k, ℓ). Notice that $m \le 2^{2n}$. Thus, we can partition $(f(k), f(\ell))$ into m non-empty sets. Fix some such partition, and let $g_{k,\ell}$ be any bijection of the equivalence classes of $\simeq$ in (k, ℓ) to the sets of the partition.

Assume, now, that k is the largest integer in M, and that $\simeq$ has m equivalence classes in the gap (k, ∞). Again, $m \leq 2^{2n}$, and k exists because M is finite. Each set a_i and each set $a_i^{\smile}$ is either finite or cofinite. The intersection of finitely many cofinite sets is cofinite, and of course any intersection that contains a finite factor must be finite. Thus, exactly one of the equivalence classes of $\simeq$ in (k, ∞) — call it y — will include an interval of the form $[p, \infty)$, and all other equivalence classes will be finite. Let k' be the largest integer in M'; of course, k' exists because M' is finite. Partition (k', ∞) into the m sets: $\{k'+i\}$ for $1 \leq i < m$, and $[k'+m, \infty)$. Take $g_{k,\infty}$ to be a bijection of the equivalence classes of $\simeq$ in (k, ∞) to the given partition of (k', ∞) such that $g_{k,\infty}(y) = [k'+m, \infty)$. Notice that $(s, \infty) \subseteq g_{k,\infty}(y)$, since

$$k' + m \leq \max(M' \sim \{\infty\}) + 2^{2n} = s\,.$$

We now define a'. Fix $i < n$, and let $p' \in Z$ be positive. We put p' and $-p'$ into a_i' according to the following stipulations.

(5) Suppose that $p' \in M'$. Then $p' \in a_i'$ iff $f^{-1}(p') \in a_i$, and $-p' \in a_i'$ iff $f^{-1}(p') \in a_i^{\smile}$.

(6) Suppose that $p' \notin M'$. Let (k, ℓ) be the (unique) positive gap in M such that $p' \in (f(k), f(\ell))$.

(a) If k, ℓ are integers and $f(\ell) - f(k) < 2^{2n}$, then $p' \in a_i'$ iff $g_{k,\ell}^{-1}(p') \in a_i$, and $-p' \in a_i'$ iff $-g_{k,\ell}^{-1}(p') \in a_i$.

(b) If k, ℓ are integers and $f(\ell) - f(k) \geq 2^{2n}$, or if $\ell = \infty$, take b and c to be the unique constituents of a and $a^{\smile}$ such that, setting $x = b \cap c \cap (k, \ell)$, we have $p' \in g_{k,\ell}(x)$. Then $p' \in a_i'$ iff a_i occurs positively in b, and $-p' \in a_i'$ iff $a_i^{\smile}$ occurs positively in c.

With the help of (6a), we establish:

(7) Let b, c be constituents of a, $a^{\smile}$, respectively, and b', c' the corresponding constituents of a', $a'^{\smile}$. Further, let (k, ℓ) be a positive gap in M. Set

$$x = b \cap c \cap (k, \ell) \quad \text{and} \quad x' = b' \cap c' \cap (f(k), f(\ell))\,.$$

(a) Assume $f(\ell) - f(k) < 2^{2n}$. Then for any p in (k, ℓ), we have $p \in x$ iff $g_{k,\ell}(p) \in x'$.

(b) Assume $f(\ell) - f(k) \geq 2^{2n}$. Then $x \neq \varnothing$ iff $x' \neq \varnothing$. Moreover, if $x \neq \varnothing$, then $g_{k,\ell}(x) = x'$.

We begin with the proof of (7a). Suppose that p is in $b \cap c \cap (k, \ell)$. Taking $g_{k,\ell}(p)$ for p' in (6a), we get

$$\begin{aligned} g_{k,\ell}(p) \in a_i' \quad &\text{iff} \quad p \in a_i && \text{by (6a)},\\ &\text{iff} \quad a_i \text{ occurs positively in } b && \text{since } p \in b,\\ &\text{iff} \quad a_i' \text{ occurs positively in } b' && \text{by definition of } b'. \end{aligned}$$

But b' is a constituent of a'. By the definition of constituent, we see that for any integer p', the following two assertions are equivalent:

$$\text{for all } i < n, \quad p' \in a'_i \quad \text{iff} \quad a'_i \text{ occurs positively in } b' ,$$
$$p' \in b' .$$

Taking $g_{k,\ell}(p)$ for p' and comparing the two previous observations, we conclude that $g_{k,\ell}(p) \in b'$. Similarly,

$$\begin{aligned} g_{k,\ell}(p) \in a'^{\smile}_i \quad &\text{iff} \quad -g_{k,\ell}(p) \in a'_i && \text{by definition of } {}^{\smile}, \\ &\text{iff} \quad -p \in a_i && \text{by (6a)}, \\ &\text{iff} \quad p \in a^{\smile}_i && \text{by definition of } {}^{\smile}, \\ &\text{iff} \quad a^{\smile}_i \text{ occurs positively in } c && \text{since } p \in c, \\ &\text{iff} \quad a'^{\smile}_i \text{ occurs positively in } c' && \text{by definition of } c'. \end{aligned}$$

Arguing as before, we readily conclude that $g_{k,\ell}(p) \in c'$. Since $g_{k,\ell}$ maps (k, ℓ) bijectively to $(f(k), f(\ell))$, we see that $g_{k,\ell}(p)$ is in $b' \cap c' \cap (f(k), f(\ell))$. This gives us the implication from left to right in (7a). Reversing the argument, we obtain the reverse implication.

We now take up the proof of (7b). Suppose that $x \neq \varnothing$. We shall prove that $x' \neq \varnothing$ and $g_{k,\ell}(x) \subseteq x'$. Certainly, $g_{k,\ell}(x) \neq \varnothing$, by definition of $g_{k,\ell}$. By using, stepwise, the definition of ${}^{\smile}$, the fact that c'corresponds to c, and the definiton of a constituent, we obtain the equivalence of the following four assertions:

$$\begin{aligned} &\text{for all } i < n, \quad -p' \in a'_i \quad \text{iff} \quad a^{\smile}_i \text{ occurs positively in } c , \\ &\text{for all } i < n, \quad p' \in a'^{\smile}_i \quad \text{iff} \quad a^{\smile}_i \text{ occurs positively in } c , \\ &\text{for all } i < n, \quad p' \in a'^{\smile}_i \quad \text{iff} \quad a'^{\smile}_i \text{ occurs positively in } c' , \\ &p' \in c' . \end{aligned}$$

Suppose that $p' \in g_{k,\ell}(x)$. Then the first assertion holds, by the definition of a' in (6b). Therefore, $p' \in c'$. A similar argument also shows that $p' \in b'$. Since $g_{k,\ell}(x) \subseteq (f(k), f(\ell))$, we conclude that $p' \in x'$. Thus $g_{k,\ell}(x) \subseteq x'$, and $x' \neq \varnothing$.

Assume, now, that $x' \neq \varnothing$. We shall show that $x \neq \varnothing$ and $x' \subseteq g_{k,\ell}(x)$. Let $p' \in x'$. Then $p' \in (f(k), f(\ell))$, by definition of x'. The range of $g_{k,\ell}$ is a partition of $(f(k), f(\ell))$, and its domain is the set of equivalence classes of $\simeq$ in (k, ℓ). Thus, there are constituents d, e of a, $a^{\smile}$, respectively, such that, setting $z = d \cap e \cap (k, \ell)$, we have $z \neq \varnothing$ and $p' \in g_{k,\ell}(z)$. However, we just finished showing that this implies p' is in $d' \cap e' \cap (f(k), f(\ell))$, where d' and e' are the constituents of a' and $a'^{\smile}$ corresponding to d and e. In particular, $p' \in d'$. Since $p' \in x'$, we also have $p' \in b'$, by definition of x'. Thus, b' and d' are constituents of a' that are not disjoint. Hence, they must be equal. Similarly, $c' = e'$. But then $b = d$ and $c = e$. Therefore, $x = z$, so $x \neq \varnothing$ and $p' \in g_{k,\ell}(x)$. Because $p' \in x'$ was arbitrary, we conclude that $x' \subseteq g_{k,\ell}(x)$. This completes the proof of (7b), and hence of (7).

Now take b to be the unique constituent of a that is cofinite. Obviously,

(8) For each $i < n$, the set a_i is cofinite iff a_i occurs positively in b.

Of course, $b^\smile$ is the corresponding cofinite constituent of $a^\smile$, so $b \cap b^\smile$ is cofinite. Put $k = \max(M \sim \{\infty\})$. The set $y = b \cap b^\smile \cap (k, \infty)$ is certainly infinite, so it is the unique infinite equivalence class of $\simeq$ in (k, ∞). Therefore, by definition of $g_{k,\infty}$, we obtain $(s, \infty) \subseteq g_{k,\infty}(y)$.

Together with (7b), this shows that

(9) $(s, \infty) \subseteq b' \cap b'^\smile \cap (f(k), \infty)$, where b' is the constituent of a' corresponding to b.

Applying (8) and (9), we get

$$\begin{aligned}(s,\infty) \subseteq a'_i \quad &\text{iff} \quad a'_i \text{ occurs positively in } b' \\ &\text{iff} \quad a_i \text{ occurs positively in } b \\ &\text{iff} \quad a_i \text{ is cofinite} \\ &\text{iff} \quad N(a_i) = -\infty \text{ and } P(a_i) = \infty\,.\end{aligned}$$

Similarly,

$$\begin{aligned}(-\infty,-s) \subseteq a'_i \quad &\text{iff} \quad (s,\infty) \subseteq a'^\smile_i \\ &\text{iff} \quad a'^\smile_i \text{ occurs positively in } b'^\smile \\ &\text{iff} \quad a^\smile_i \text{ occurs positively in } b^\smile \\ &\text{iff} \quad a_i \text{ is cofinite.}\end{aligned}$$

In a completely analogous way, we show that

$$\begin{aligned}(-\infty,-s) \text{ and } (s,\infty) \text{ are disjoint from } a'_i \quad &\text{iff} \quad a_i \text{ occurs negatively in } b \\ &\text{iff} \quad a_i \text{ is finite} \\ &\text{iff} \quad -\infty < N(a_i) \text{ and } P(a_i) < \infty.\end{aligned}$$

This verifies that a' is a sequence of finite and cofinite sets of integers, since a is certainly such a sequence. Therefore, $a' \in {}^nA$. It also shows that (i) holds. Further, it shows that a and a' satisfy property (P3) in Lemma 4.10.

To establish (ii) it suffices to verify that a and a' satisfy the hypothesis of Lemma 4.10. We have already verified condition (P3). It remains to check (P1) and (P2). Condition (P1) follows at once from (7). To check condition (P2), suppose that $p \in M$. Then $f(p) \in M'$. By the stipulations of (5) (with "$f(p)$" in place of "p'") and the fact that f preserves $-$, we have $p \in a_i$ iff $f(p) \in a'_i$. Thus, (P2) holds for a and a'. This completes the proof of the lemma. □

We shall need the following number-theoretic fact in the proof of the next lemma.

LEMMA 4.12. *For all integers k, ℓ, m with $k > 0$, $\ell \geq 0$, and $m \geq 2$, there is an integer k' in $(\ell, \ell + m!]$ such that $k' \equiv k \mod d$ for every $d = 2, \ldots, m$.*

PROOF. Obviously,

$$m! \equiv 0 \mod d \quad \text{for} \quad d = 2, \ldots, m\,.$$

Therefore, for any $q \in Z$,

$$k + q \cdot m! \equiv k \mod d \quad \text{for} \quad d = 2, \ldots, m\,.$$

Now two contiguous numbers in the arithmetic sequence $\langle k + q \cdot m! : q \in Z \rangle$ are separated by a distance of $m!$. Hence, any interval of length $m!$ must contain one of the numbers in the sequence. In particular, $(\ell, \ell + m!]$ contains one of these numbers. We take it to be k'. □

For a given integer $m \geq 1$, let R_m denote the $2m$-ary relation on Z that holds for $(x_0, \ldots, x_{m-1}, y_0, \ldots, y_{m-1})$ iff $\sum_{i<m} x_i = \sum_{i<m} y_i$. Of course, R_m is definable in $\mathfrak{Z}_e$ using the ternary relation $+$. However, it may not be definable in a substructure $\mathfrak{M} \subseteq \mathfrak{Z}_e$: even though $x_0, \ldots, x_{m-1}, y_0, \ldots, y_{m-1}$ are in M, some, or all, of the individual subsums (e.g., $x_0 + x_1$, $x_0 + x_1 + x_2$, etc.) may fail to be in M. Similar remarks hold for the $2m$-ary relation S_m expressing the fact that $\sum_{i<m} x_i < \sum_{i<m} y_i$, and for the m-ary relations T_d (one for each integer $d > 1$) expressing the fact that d divides $\sum_{i<m} x_i$. It should not cause any confusion if we continue to use traditional notation and write

$$\begin{aligned} \textstyle\sum_{i<m} x_i &= \textstyle\sum_{i<m} y_i\,, \\ \textstyle\sum_{i<m} x_i &< \textstyle\sum_{i<m} y_i\,, \\ d &\mid \textstyle\sum_{i<m} x_i\,, \end{aligned}$$

instead of

$$\begin{aligned} &R_m(x_0, \ldots, x_{m-1}, y_0, \ldots, y_{m-1})\,, \\ &S_m(x_0, \ldots, x_{m-1}, y_0, \ldots, y_{m-1})\,, \\ &T_d(x_0, \ldots, x_{m-1})\,. \end{aligned}$$

The reader should keep in mind that when this notation is used, only the elements $x_0, \ldots, x_{m-1}, y_0, \ldots, y_{m-1}$ (not d or the individual subsums) are assumed to be in the structure $\mathfrak{M}$ under discussion.

DEFINITION 4.13. An isomorphism f between substructures $\mathfrak{M}$ and $\mathfrak{M}'$ of $\mathfrak{Z}_e$ *preserves equality of m-sums*, *inequality of m-sums*, or *divisibility of m-sums by d* according as, for all $x_0, \ldots, x_{m-1}, y_0, \ldots, y_{m-1}$ in M,

$$\begin{aligned} \textstyle\sum_{i<m} x_i = \sum_{i<m} y_i \quad &\text{iff} \quad \textstyle\sum_{i<m} f(x_i) = \sum_{i<m} f(y_i)\,, \\ \textstyle\sum_{i<m} x_i < \sum_{i<m} y_i \quad &\text{iff} \quad \textstyle\sum_{i<m} f(x_i) < \sum_{i<m} f(y_i)\,, \end{aligned}$$

or

$$d \mid \sum_{i<m} x_i \quad \text{iff} \quad d \mid \sum_{i<m} f(x_i)\,,$$

respectively. □

REMARK. Notice that, for $1 \leq k \leq m$, every k-sum is an m-sum. This follows at once from the fact that 0 is a distinguished element of $\mathfrak{M}$ and therefore may be used to write a k-sum as an m-sum. Hence, if f preserves, e.g., the equality of m-sums, then it must preserve the equality of a k-sum with an ℓ-sum whenever $1 \leq k, \ell \leq m$. Similar remarks concerning the preservation of the inequality of a k-sum with an ℓ-sum, and of the divisibility of k-sums by d hold as well. □

LEMMA 4.14 (ISOMORPHISM EXTENSION LEMMA). *Let f be an isomorphism between two finite substructures $\mathfrak{M}$ and $\mathfrak{M}'$ of $\mathfrak{Z}_e$. Further, for k a positive integer, set $\overline{k} = 4k^2 \cdot (k! + 2)$, and suppose that f preserves the equality and inequality of $\overline{k}$-sums and the divisibility of $\overline{k}$-sums by all d satisfying $1 < d \leq \overline{k}$. Then for any integer $b > \max(M - \{\infty\})$ there is an integer b' with*

$$\max(M' {\sim} \{\infty\}) < b' \leq 2k \cdot \max(M' {\sim} \{\infty\}) + k!$$

such that the substructures of $\mathfrak{Z}_e$ obtained by adding $b, -b$ to $\mathfrak{M}$ and $b', -b'$ to $\mathfrak{M}'$ are isomorphic via the mapping f_1 that extends f and that sends b to b' and $-b$ to $-b'$. Moreover, f_1 preserves the equality and inequality of k-sums and the divisibility of k-sums by all d such that $1 < d \leq k$.

PROOF. Assume that $\mathfrak{M}$, $\mathfrak{M}'$, f, k, and $\overline{k}$ satisfy the hypotheses of the lemma. The choice of b' is somewhat delicate. We shall need to investigate certain quotients of $2k$-sums from $\mathfrak{M}$ (in fact, rational numbers of the form $(\sum_{i<2k} x_i)/n$ where $x_0, \ldots, x_{2k-1}$ are integers from $\mathfrak{M}$ and $1 \leq n \leq 2k$) and establish the preservation under f of various relationships between these quotients. The Σ-notation that we shall employ will involve either k-sums or $2k$-sums, and we shall always specify which. Thus, the index set of summation will be unambiguously determined. To simplify notation we shall write, e.g., "$\sum x_i$" instead of "$\sum_{i<2k} x_i$". Also, for each integer $x \in M$ we set $x' = f(x)$; thus, we shall write "x'" in place of "$f(x)$".

(1) For any integers $x_0, \ldots, x_{2k-1}, y_0, \ldots, y_{2k-1}$ in M and any positive integers $m, n \leq 2k$, we have

$$\begin{aligned} \left(\textstyle\sum x_i\right)/n < \left(\textstyle\sum y_j\right)/m \quad &\text{iff} \quad \left(\textstyle\sum x_i'\right)/n < \left(\textstyle\sum y_j'\right)/m\,, \\ \left(\textstyle\sum x_i\right)/n = \left(\textstyle\sum y_j\right)/m \quad &\text{iff} \quad \left(\textstyle\sum x_i'\right)/n = \left(\textstyle\sum y_j'\right)/m\,. \end{aligned}$$

To prove (1), notice that $4k^2 \leq \overline{k}$. Thus, f preserves $4k^2$-inequalities. Notice also that $m \cdot (\sum x_i)$ and $n \cdot (\sum y_i)$ are really just $4k^2$-sums; e.g.,

$$m \cdot \left(\textstyle\sum x_i\right) = \textstyle\sum_{q<m} \left(\sum x_i\right).$$

Using these facts and elementary arithmetic, we obtain

$$\begin{aligned} \left(\sum x_i\right)/n < \left(\sum y_j\right)/m \quad &\text{iff} \quad m \cdot \left(\sum x_i\right) < n \cdot \left(\sum y_j\right) \\ &\text{iff} \quad m \cdot \left(\sum x_i'\right) < n \cdot \left(\sum y_j'\right) \\ &\text{iff} \quad \left(\sum x_i'\right)/n < \left(\sum y_j'\right)/m\,. \end{aligned}$$

The proof that equality is preserved is obtained by replacing "$<$" everywhere with "$=$".

For any integer x and any integer $m > 1$, let $r_m(x)$ denote the remainder of x upon division by m. Thus, $r_m(x)$ is the unique integer r such that $0 \leq r < m$ and, for some integer t, we have $x = tm + r$.

(2) For any integers $x_0, \ldots, x_{2k-1}$ in M and any integer d with $1 < d \leq 2k$,

$$r_d\left(\sum x_i\right) = r_d\left(\sum x_i'\right)\,.$$

To prove (2), set $r = r_d\left(\sum x_i\right)$, say $\left(\sum x_i\right) = dt + r$ where t is some integer. Notice that $\left(\sum x_i\right) - r$ is a $4k$-sum — and hence a $\overline{k}$-sum — from $\mathfrak{M}$, since $0 \leq r < 2k$ and since $-r$ can be written as -1 added to itself r times. (Recall here that -1 is in M.) Notice also that $\left(\sum x_i\right) - r$ is divisible by d and that $d \leq \overline{k}$. By assumption, f preserves the divisibility of $\overline{k}$-sums by such d. Also, $f(-1) = -1$. Therefore $\left(\sum x_i'\right) - r$ is divisible by d, i.e., $\sum x_i' - r = dp$ for some integer p. But then r is the remainder of $\sum x_i'$ upon division by d, which proves (2).

The next step in the proof is the key to the proper choice of b'. Roughly speaking, it says that the integer distance between restricted quotients of $2k$-sums in M is preserved by f, up to a distance of $k!$. The computations in this step are a decisive factor in determining the size that we specified for $\overline{k}$ in relationship to k. For any rational number q, let $\lceil q \rceil$ be the least integer greater than or equal to q.

(3) For any integers $x_0, \ldots, x_{2k-1}, y_0, \ldots, y_{2k-1}$ in M and any positive integers $m, n \leq 2k$ such that $0 \leq \left(\sum x_i\right)/n \leq \left(\sum y_j\right)/m$, the integers

$$\left\lceil \left(\sum y_j\right)/m \right\rceil - \left\lceil \left(\sum x_i\right)/n \right\rceil \quad \text{and} \quad \left\lceil \left(\sum y_j'\right)/m \right\rceil - \left\lceil \left(\sum x_i'\right)/n \right\rceil$$

are either equal to each other or both greater than $k!$.

To prove (3), set

$$\begin{aligned} d &= \left\lceil \left(\sum y_j\right)/m \right\rceil - \left\lceil \left(\sum x_i\right)/n \right\rceil, \\ e &= \left\lceil \left(\sum y_j'\right)/m \right\rceil - \left\lceil \left(\sum x_i'\right)/n \right\rceil, \end{aligned}$$

$$s = \begin{cases} n - r_n\left(\sum x_i\right) & \text{if } n \nmid \sum x_i \\ 0 & \text{if } n \mid \sum x_i \end{cases},$$

$$t = \begin{cases} m - r_m\left(\sum y_j\right) & \text{if } m \nmid \sum y_j \\ 0 & \text{if } m \mid \sum y_j \end{cases}.$$

Using these definitions, simple computations yield

$$\begin{aligned}\lceil(\sum x_i)/n\rceil &= (\sum x_i + s)/n\,,\\ \lceil(\sum y_j)/m\rceil &= (\sum y_j + t)/m\,,\\ \lceil(\sum x_i)/n\rceil + d &= \lceil(\sum y_j)/m\rceil\,,\\ (\sum x_i + s)/n + d &= (\sum y_j + t)/m\,,\end{aligned}$$

and hence

$$m\cdot(\sum x_i) + ms + mnd = n\cdot(\sum y_j) + nt\,. \tag{4}$$

Using the definitions of s and t, we see from (2) and from the assumption that the divisibilty of $\overline{k}$-sums is preserved, that

$$s = \begin{cases} n - r_n(\sum x'_i) & \text{if } n \nmid \sum x'_i \\ 0 & \text{if } n \mid \sum x'_i \end{cases},$$

$$t = \begin{cases} m - r_m(\sum y'_j) & \text{if } m \nmid \sum y'_j \\ 0 & \text{if } m \mid \sum y'_j \end{cases}.$$

Therefore, a computation similar to the preceding one yields

$$\begin{aligned}\lceil(\sum x'_i)/n\rceil &= (\sum x'_i + s)/n\,,\\ \lceil(\sum y'_j)/m\rceil &= (\sum y'_j + t)/m\,,\\ (\sum x'_i + s)/n + e &= (\sum y'_j + t)/m\,,\end{aligned}$$

and hence

$$m\cdot(\sum x'_i) + ms + mne = n\cdot(\sum y'_j) + nt\,. \tag{5}$$

Observe that $0 \le (\sum x'_i)/n \le (\sum y'_j)/m$, by (1) and the preservation properties of f.

Suppose now that $d \le k!$. Then mnd can be written as a $4k^2k!$-sum (where the number of summands that are 1 is mnd, and the rest of the summands are 0). Also,

$$m\cdot(\sum x_i) \quad , \quad n\cdot(\sum y_j) \quad , \quad ms \quad , \quad nt$$

are $4k^2$-sums, since $\sum x_i$ and $\sum y_j$ are $2k$-sums and we have

$$1 \le m, n \le 2k \quad , \quad 0 \le s, t \le 2k\,.$$

Therefore, the left-hand side of (4) is a sum of $\overline{k}$ (i.e., $4k^2 + 4k^2 + 4k^2k!$) integers from M. The right-hand side is a sum of $4k^2 + 4k^2$ — and hence also of $\overline{k}$ — integers from M. Because f preserves the equality of $\overline{k}$-sums, and also preserves 0 and 1, we obtain

$$m\cdot(\sum x'_i) + ms + mnd = n\cdot(\sum y'_j) + nt\,. \tag{6}$$

Comparing (5) with (6), we see that we must have $d = e$.

If $e \leq k!$, then the same argument applied to (5) leads to (4), with "d" replaced by "e". Hence, we again obtain $d = e$. This proves (3).

Set $M_1 = M \cup \{-b, b\}$. Fix an arbitrary integer $b' > \max(M' \sim \{\infty\})$, and set

$$M_1' = M' \cup \{-b', b'\} \quad , \quad f_1 = f \cup \{(b, b'), (-b, -b')\}\,.$$

We establish conditions under which f_1 will preserve k-sums and will preserve divisibility of k-sums by positive integers $d \leq k$.

(7) Suppose that for any integers $x_0, \ldots, x_{2k-1}$ in M and any positive integer $n \leq 2k$ we have

$$\left(\sum x_i\right)/n < b \quad \text{iff} \quad \left(\sum x_i'\right)/n < b'\,,$$
$$\left(\sum x_i\right)/n = b \quad \text{iff} \quad \left(\sum x_i'\right)/n = b'\,.$$

Then f_1 preserves the equality and inequality of k-sums of M_1.

To prove (7), assume that $nb + \sum x_i$ and $mb + \sum y_j$ are two k-sums of M_1 (thus, b or $-b$ is a summand $|n|$ times in the first sum and $|m|$ times in the second sum, x_i and y_j are integers in M for $i < k - |n|$ and $j < k - |m|$, and of course $-k \leq m, n \leq k$), and suppose that

(8) $\quad nb + \sum x_i < mb + \sum y_j\,.$

If $m = n$, then $\sum x_i < \sum y_j$. Since f preserves the inequality of $\bar{k}$-sums in $\mathfrak{M}$, we get $\sum x_i' < \sum y_j'$ and hence

(9) $\quad nb' + \sum x_i' < mb' + \sum y_j'\,,$

as desired. If $m > n$, then

$$\sum x_i - \sum y_j < (m - n)b\,,$$

so

$$\left(\sum x_i - \sum y_j\right)/(m - n) < b\,.$$

Now $\sum x_i - \sum y_j$ is certainly a $2k$-sum, since each summand is a k-sum. Therefore, we can apply the hypotheses of (7) to conclude that

$$\left(\sum x_i' - \sum y_j'\right)/(m - n) < b'\,.$$

Using elementary arithmetic, we eventually arrive at (9). The proof when $m < n$ is similar. We use the fact that, under the hypotheses of (7), we also have

$$\left(\sum x_i\right)/n > b \quad \text{iff} \quad \left(\sum x_i'\right)/n > b'\,.$$

We have shown that (8) implies (9). By running the argument in reverse, we see that (9) implies (8). Thus, (8) and (9) are equivalent. The proof that (8) and (9) remain equivalent when "$<$" is replaced by "$=$" is similar. This completes the proof of (7).

(10) Suppose that $b \equiv b' \mod d$ for all integers d with $1 < d \leq k$. Then f_1 preserves divisibility of k-sums of M_1 by each d with $1 < d \leq k$.

Indeed, to assume that $b \equiv b' \mod d$ is to assume that

(11) $r_d(b) = r_d(b')$.

Let $nb + \sum x_i$ be a k-sum of M_1 (with $x_i \in M$ for $i < k - |n|$ and $-k \leq n \leq k$). Using elementary arithmetic we obtain

$$\begin{aligned} r_d\left(nb + \textstyle\sum x_i\right) &= r_d\left(n \cdot r_d(b) + r_d\left(\textstyle\sum x_i\right)\right) \\ &= r_d\left(n \cdot r_d(b') + r_d\left(\textstyle\sum x_i'\right)\right) \quad \text{by (2), (11),} \\ &= r_d\left(nb' + \textstyle\sum x_i'\right) . \end{aligned}$$

Thus, d divides $nb + \sum x_i$ iff it divides $nb' + \sum x_i'$. This proves (10).

We are ready to choose b'. Let p be the largest rational number of the form $\left(\sum_{i<2k} z_i\right)/m$ that is less than or equal to b, where $z_0, \ldots, z_{2k-1}$ range over integers in M and $1 \leq m \leq 2k$. Such a maximum certainly exists and in fact is positive, since M is finite, 0 and 1 are in M, and $b \geq 1$. (The set of rational numbers that can be written in the prescribed form is finite and contains 1. Therefore, $p \geq 1$.) Let q be the smallest rational number of the form $\left(\sum_{i<2k} z_i\right)/m$ that is greater than or equal to b (where $z_0, \ldots, z_{2k-1}$ and m are as above), provided that such a number exists at all; if it does not exist, set $q = \infty$. Suppose $p = \left(\sum y_j\right)/m$. Certainly, $\sum y_j > 0$, since $p > 0$.

CASE 1: q is finite and $\lceil q \rceil - \lceil p \rceil \leq k!$.

Since $p \leq b \leq q$, there is an e with $0 \leq e \leq k!$ such that $b = \lceil (\sum y_j)/m \rceil + e$. Set

$$b' = \lceil (\textstyle\sum y_j')/m \rceil + e .$$

We first show that f_1 (as defined before (7), but for the b' just specified) preserves the equality and inequality of k-sums of M_1. By (7), it suffices to show that b' satisfies the hypotheses of (7), or, equivalently, that

$$\begin{aligned} (\textstyle\sum x_i)/n \leq b \quad &\text{iff} \quad (\textstyle\sum x_i')/n \leq b' , \\ (\textstyle\sum x_i)/n \geq b \quad &\text{iff} \quad (\textstyle\sum x_i')/n \geq b' . \end{aligned} \tag{12}$$

Let $\sum x_i$ be a $2k$-sum from M and n an integer between 1 and $2k$. Then

$$\begin{aligned} (\textstyle\sum x_i)/n \leq b \quad &\text{iff} \quad (\textstyle\sum x_i)/n \leq (\textstyle\sum y_j)/m \quad \text{by the maximality of } p, \\ &\text{iff} \quad (\textstyle\sum x_i')/n \leq (\textstyle\sum y_j')/m \quad \text{by (1).} \end{aligned} \tag{13}$$

We shall show that these conditions are equivalent to the condition $(\sum x_i')/n \leq b'$. Certainly, if $(\sum x_i')/n \leq (\sum y_j')/m$, then $(\sum x_i')/n \leq b'$, since $e \geq 0$. For the reverse implication, suppose that $(\sum x'_i)/n \le b'$. Assume, for contradiction, that $(\sum x'_i)/n > (\sum y'_j)/m$. Since $\sum y_j$ is positive, $\sum y'_j$ is also positive, by (1). We therefore have

$$0 < \lceil (\sum y'_j)/m \rceil \le \lceil (\sum x'_i)/n \rceil \le b' = \lceil (\sum y'_j)/m \rceil + e\,.$$

Hence,

$$0 \le \lceil (\sum x'_i)/n \rceil - \lceil (\sum y'_j)/m \rceil \le e \le k!\,.$$

By (3), this gives

$$0 \le \lceil (\sum x_i)/n \rceil - \lceil (\sum y_j)/m \rceil \le e\,,$$

so

$$0 < \lceil (\sum x_i)/n \rceil \le \lceil (\sum y_j)/m \rceil + e = b\,.$$

Thus, $(\sum x_i)/n \le b$. By (13), this forces $(\sum x'_i)/n \le (\sum y'_j)/m$, which is the desired contradiction. This establishes the first equivalence in (12). The second is established in a completely analogous fashion.

To check that f_1 preserves the divisibility of k-sums of M_1 by d whenever the inequalities $1 < d \le k$ hold, we must verify the conditions of (10). Let d be given, and set $t = r_d(b)$. Our goal is to prove that $t = r_d(b')$. Put

$$s = \begin{cases} m - r_m(\sum y_j)/m & \text{if } m \nmid \sum y_j \\ 0 & \text{if } m \mid \sum y_j \end{cases}.$$

By (2), we have

$$r_m(\sum y_j) = r_m(\sum y'_j)\,,$$

and, in particular,

$$m \mid (\sum y_j) \quad \text{iff} \quad m \mid (\sum y'_j)\,.$$

Thus,

$$s = \begin{cases} m - r_m(\sum y'_j)/m & \text{if } m \nmid \sum y'_j \\ 0 & \text{if } m \mid \sum y'_j \end{cases}.$$

Therefore, using the definitions of b, b', and s, we compute that

(14) $\quad b = (\sum y_j + s)/m + e \quad$ and $\quad b' = (\sum y'_j + s)/m + e\,.$

Now $b = du + t$ for some integer u and some t with $0 \le t < d$, by definition of $r_d(b)$. Combining this with (14) and using some simple arithmetic, we arrive at

(15) $\quad \sum y_j + s + me - mt = dmu\,.$

Thus, the left-hand side of (15) is divisible by dm. Now $1 \le dm \le 2k^2 \le \overline{k}$. Also, since $1 \le m \le 2k$, the summand $\sum y_k$ is a $2k$-sum, s is a $2k$-sum, me is a $2k \cdot k!$-sum (of ones and zeros), and $-mt$ is a $2k^2$-sum (of ones and zeros). Therefore, the sum on the left-hand side of (15) is a $\overline{k}$- sum from M. Because f is assumed to preserve the divisibility of $\overline{k}$-sums by positive integers $\le \overline{k}$, there must be an integer v such that

$$\sum y'_j + s + me - mt = dmv\,.$$

Consequently,

$$\left(\sum y'_j + s\right)/m + e = dv + t\,,$$

i.e., $b' = dv + t$, by (14). This completes the proof that $t = r_d(b')$.

It remains to show that b' falls within the desired interval. Let

$$c = \max(M \sim \{\infty\}) \quad \text{and} \quad c' = \max(M' \sim \{\infty\})\,.$$

Since f preserves $<$, it must map c to c'. Because $b > c$, and because f_1 preserves the equality and inequality of k-sums of M_1, we get $b' > c'$. Because $y'_j \le c'$ for each j, we certainly have $\sum y'_j \le 2kc'$ and hence $(\sum y'_j)/m \le 2kc'$. Therefore,

$$b' = \lceil \left(\sum y'_j\right)/m \rceil + e \le 2kc' + e \le 2kc' + k!\,.$$

CASE 2: Either $q = \infty$ or $\lceil q \rceil - \lceil p \rceil > k!$.

Set $\ell = \lceil \left(\sum y'_j\right)/m \rceil$. Using Lemma 4.12, we may choose a b' in $(\ell, \ell + k!]$ such that $b' \equiv b \mod d$ for every d with $1 < d \le k$. Hence, by (10), f_1 preserves the divisibility of k-sums of M_1 for each such d.

To establish that f_1 preserves the equality and inequality of k-sums of M_1 we must again verify the conditions of (7). We derive (13) just as before. Also as before, the last inequality in (13) implies that $\left(\sum x'_i\right)/n \le b'$. To establish the reverse implication, we argue by contraposition. Consider first the case when q is finite, and suppose that $\left(\sum x'_i\right)/n > \left(\sum y'_j\right)/m$. Proceeding stepwise, we obtain:

$\left(\sum x_i\right)/n > \left(\sum y_j\right)/m$	by (1),
$\left(\sum x_i\right)/n \ge q$	by the maximality of p and the minimality of q,
$\lceil\left(\sum x_i\right)/n\rceil - \lceil\left(\sum y_j\right)/m\rceil > k!$	by the definition of p and the assumption $\lceil q \rceil - \lceil p \rceil > k!$,
$\lceil\left(\sum x'_i\right)/n\rceil - \lceil\left(\sum y'_j\right)/m\rceil > k!$	by (3),
$\lceil\left(\sum x'_i\right)/n\rceil > \lceil\left(\sum y'_j\right)/m\rceil + k! = \ell + k!$,	
$\lceil\left(\sum x'_i\right)/n\rceil > b'$	by choice of b'.

The proof of the reverse implication when $q = \infty$ is trivial, since in this case there can be no $\left(\sum x_i\right)/n$ that is greater than $\left(\sum y_j\right)/m$. The verification that $(\sum x_i)/n \ge b$ iff $(\sum x'_i)/n \ge b'$ is completely similar.

Finally, $b' > \max(M' \sim \{\infty\})$, just as in the previous case. Since

$$\ell \le 2k \cdot \max(M' \sim \{\infty\})$$

(as before), we get $b' \leq 2k \cdot \max(M' \sim \{\infty\})$ by the choice of b'. This completes the proof of lemma. □

We define two sequences of recursive functions,

$$\langle g_n : 2 \leq n \leq \omega \rangle \quad \text{and} \quad \langle h_{m,n} : 2 \leq n < \omega \text{ and } 0 \leq m \leq n \rangle .$$

The domain of g_n is ω, while the domain of $h_{m,n}$ is $\{0, \ldots, m\}$. The functions are determined by the stipulations:

$$\begin{aligned} g_n(0) &= n\,, \\ g_n(i) &= 4g_n(i-1)^2 \cdot [g_n(i-1)! + 2] \quad \text{for} \quad 0 < i\,; \\ h_{m,n}(0) &= 1\,, \\ h_{m,n}(i) &= 2g_n(m-i) \cdot h_{m,n}(i-1) + g_n(m-i)! \quad \text{for} \quad 0 < i \leq m\,. \end{aligned}$$

In understanding the formulation of the next lemma it is helpful to observe that each finite substructure $\mathfrak{M}$ of $\mathfrak{Z}_e$ has cardinality $2m+5$, where m is the number of integers in $\mathfrak{M}$ that are greater than 1. Indeed, the five elements $-\infty$, -1, 0, 1, ∞ are in $\mathfrak{M}$, and for each b in $\mathfrak{M}$ different from these five elements, $-b$ is also in $\mathfrak{M}$.

COROLLARY 4.15. *Let $\mathfrak{M}$ be a finite substructure of $\mathfrak{Z}_e$ of cardinality $2m+5$, and n any integer ≥ 2. Then there is a substructure $\mathfrak{M}'$ of $\mathfrak{Z}_e$ and a function f from M to M' such that*:

(i) *f is an isomorphism between $\mathfrak{M}$ and $\mathfrak{M}'$*;
(ii) *f preserves the equality and inequality of n-sums and the divisibility of n-sums by d whenever $1 < d \leq n$*;
(iii) $\max(M' - \{\infty\}) \leq h_{m,n}(m)$.

PROOF. Let $\mathfrak{M}$ be given, and let $b_1 < b_2 < \cdots < b_m$ be an enumeration of the integers in M that are greater than 1. For each $i = 0, \ldots, m$, set

$$M_i = \{-\infty, -1, 0, 1, \infty\} \cup \{b_\ell : 1 \leq \ell \leq i\} \cup \{-b_\ell : 1 \leq \ell \leq i\}\,,$$

and let $\mathfrak{M}_i$ be the corresponding substructure of $\mathfrak{Z}_e$. We proceed by induction on i to construct a chain (under $\subseteq$) of substructures $\langle \mathfrak{M}'_i : i \leq m \rangle$ of $\mathfrak{Z}_e$ and a chain (under $\subseteq$) of functions $\langle f_i : i \leq m \rangle$ such that, for each $i \leq m$,

(1) f_i maps $\mathfrak{M}_i$ isomorphically onto $\mathfrak{M}'_i$,

(2) f_i preserves the equality and inequality of $g_n(m-i)$-sums and the divisibility of $g_n(m-i)$-sums by d whenever $1 < d \leq g_n(m-i)$,

(3) $\max(M'_i - \{\infty\}) \leq h_{m,n}(i)$.

For $i = 0$, set $\mathfrak{M}'_0 = \mathfrak{M}_0$ and let f_0 be the identity function on M_0. The verification of (1)–(3) is trivial. Now suppose that $\mathfrak{M}'_i$ and f_i have been constructed satisfying (1)–(3), and suppose that $i+1 \leq m$. Set

(4) $k = g_n(m-(i+1)) = g_n(m-i-1)$,

and observe that $g_n(m-i) = 4k^2(k!+2)$. In view of the induction hypotheses (1) and (2), we can apply Lemma 4.14 to $\mathfrak{M}_i$, $\mathfrak{M}'_i$ and f_i to obtain an extension $\mathfrak{M}'_{i+1}$ of $\mathfrak{M}'_i$ and an isomorphism f_{i+1} from $\mathfrak{M}_{i+1}$ to $\mathfrak{M}'_{i+1}$ that extends f_i and that preserves the equality and inequality of k-sums and the divisibility of k-sums by d whenever $1 < d \leq k$. Thus, in view of (4), we see that (1) and (2) hold with "$i+1$" in place of "i". We verify (3) for "$i+1$" as follows:

$$\begin{aligned}\max(M'_{i+1} - \{\infty\}) &\leq 2k \cdot \max(M'_i - \{\infty\}) + k! && \text{by Lemma 4.14,}\\ &\leq 2k \cdot h_{m,n}(i) + k! && \text{by the induction hypothesis (3),}\\ &= h_{m,n}(i+1) && \text{by (4) and the definition of } h_{m,n}\,.\end{aligned}$$

This completes the construction and the verification of properties (1)–(3). Taking $\mathfrak{M}' = \mathfrak{M}'_m$ and $f = f_m$, we obtain conclusions (i)–(iii) from (1)–(3) with $i = m$. □

THEOREM 4.16. *The equational theory of $\mathfrak{A}$ is decidable.*

PROOF. Suppose $\tau(v_0, \ldots, v_{n-1})$ is any relation algebraic term such that $\tau \neq 0$ is satisfiable in $\mathfrak{A}$. Let $a \in {}^nA$ be an assignment such that $\tau[a] \neq \varnothing$, and set

$$I = \bigcup\{In_{\sigma[a]} : \sigma \quad \text{is a subterm of} \quad \tau\}\,.$$

(Recall that $In_{\sigma[a]}$ is the set of indicators of $\sigma[a]$.) Take $\mathfrak{M}$ to be the substructure of $\mathfrak{Z}_e$ with universe

$$\{-\infty, -1, 0, 1, \infty\} \cup I \cup \{-b : b \in I\}\,,$$

and suppose the cardinality of $\mathfrak{M}$ is $2m+5$. We apply Corollary 4.15 to the integer 2^{2n} (in place of n) to obtain a substructure $\mathfrak{M}'$ of $\mathfrak{Z}_e$ and a mapping f satisfying conditions (i)–(iii) of 4.15, with "n" replaced by "2^{2n}" in (ii) and (iii).

Assume that (k, ℓ) is a gap in $\mathfrak{M}$ with integer endpoints and such that

$$f(\ell) - f(k) < 2^{2n}\,.$$

Because $f(k)$, $f(\ell)$ are in M',

$$f(k) + \underbrace{1 + \cdots + 1}_{f(\ell)-f(k) \text{ times}} = f(\ell)$$

is an equality between 2^{2n}-sums in $\mathfrak{M}'$. Since f preserves such equalities, by 4.15(ii), and since f preserves 1, we obtain

$$k + \underbrace{1 + \cdots + 1}_{f(\ell)-f(k) \text{ times}} = \ell\,.$$

In other words, $\ell - k = f(\ell) - f(k)$.

We have verified that condition (P4) in Lemma 4.11 holds. Therefore, there is an assignment a' in ${}^nA$ such that the conclusions 4.11(i),(ii) hold. In particular, from 4.11(ii) and 4.7 we conclude that $\tau[a'] \neq \varnothing$. From 4.11(i) and 4.15(iii) we see that, for each $i < n$, the set a'_i is either a subset of the interval

$$[-2^{2n} - h_{m,2^{2n}}(m)\,,\, 2^{2n} + h_{m,2^{2n}}(m)]$$

or it is the union of such a subset with the complement of the interval.

Take p to be the number of subterms of τ. Then $m \leq 4p$, i.e., $|M| \leq 8p + 5$. Indeed, for any subterm σ of τ, the set $\sigma[a]$ has at most 4 indicators that are different from $-\infty$, -1, 0, 1, and ∞.

Putting the observations of the preceding paragraphs together, we arrive at the following conclusion. Suppose $\tau(v_0, \ldots, v_{n-1})$ is any relation-algebraic term, say with p subterms. If the inequality $\tau \neq 0$ is satisfiable at all in $\mathfrak{A}$, then it is satisfiable by an assignment $a' \in {}^nA$ such that, for each $i < n$, the set a'_i is either a subset of the interval

$$J = [-2^{2n} - h_{4p,2^{2n}}(4p)\,,\, 2^{2n} + h_{4p,2^{2n}}(4p)]$$

or the union of such a subset with the complement of J. Because J is finite, there are only finitely many such assignments (in fact, $(2 \cdot 2^{|J|})^n = 2^{n \cdot |J| + n}$).

Thus, to decide if $\tau \neq 0$ is satisfiable in $\mathfrak{A}$, we first compute p and $h_{4p,2^{2n}}(4p)$ to arrive at the interval J. We then construct each of the finitely many assignments a' as above, and verify whether any such a' actually satisfies $\tau \neq 0$ in $\mathfrak{A}$. Because the operations of $\mathfrak{A}$ are all recursive, such a verification is possible. This completes the proof of the theorem. □

REMARK. It is interesting to contrast the relation algebras $\mathfrak{A}$ and $\mathfrak{Cf}(3)$. Except for their relative product operations, these algebras are identical, i.e., the reducts of the two algebras obtained by eliminating relative multiplication coincide. Yet the equational theory of $\mathfrak{A}$ is decidable, by Theorem 4.16, while that of $\mathfrak{Cf}(3)$ is undecidable and in fact recursively inseparable, by Corollary 2.12. □

PROBLEM 4.17. In Jipsen-Lukács [1994] a symmetric, simple, representable relation algebra $\mathfrak{B}$ that is infinite but generated by a single element is defined as follows. Its universe is the collection $Cf(\omega)$ of the finite and cofinite subsets of natural numbers. The Boolean operations are the set-theoretic ones, conversion is the identity function on $Cf(\omega)$, the identity element is $\{0\}$, and relative multiplication operation is defined on atoms $\{m\}$ and $\{n\}$ by the rule

$$\{m\} \,;\, \{n\} = [\,|m - n|\,, m + n]\,.$$

It is shown in *op. cit.* that the only consistent equational theory of relation algebras properly including $\mathcal{E}q(\mathfrak{B})$ is the theory of the algebra of constants over a base set of cardinality 3. It seems that $\mathfrak{B}$ has several properties in common with the algebra $\mathfrak{A}$. Does $\mathfrak{B}$ have a decidable equational theory? □

PROBLEM 4.18. Let O be the trivial variety of one-element relation algebras. Is there a variety V of relation algebras such that, in the lattice of varieties of relation algebras, the interval $[\mathsf{O}, \mathsf{V}]$ is finite, yet V has an undecidable equational theory? □

HISTORICAL REMARK. The results of this chapter were obtained by the authors in 1991. They were mentioned in a lecture given by the second author in 1991 at the Banach Center, but were not announced in print until Andréka-Givant-Németi [1994a]. □

BIBLIOGRAPHY

[1995] Andréka, H. and Bredikhin, D. A., *The equational theory of union-free algebras of relations*, Algebra Universalis **33** (1995), 516–532.

[1994] Andréka, H., Givant, S. R., and Németi, I., *The lattice of varieties of representable relation algebras*, Journal of Symbolic Logic **59** (1994), 631–661.

[1994a] ———, *Decision problems for equational theories of relation algebras*, Bulletin of the Section of Logic **23** (1994), 47–52.

[1994b] ———, *Undecidable equational theories of relation algebras*, Abstracts of papers presented to the American Mathematical Society **15** (1994), 475.

[1935] Birkhoff, G., *On the structure of abstract algebras*, Proceedings of the Cambridge Philosophical Society **31** (1935), 433–454.

[1973] Chang, C. C. and Keisler, H. J., *Model theory*, Studies in Logic and the Foundations of Mathematics vol. 73, North-Holland Publishing Company, Amsterdam, 1973, xii + 550 pp.

[1948] Chin, L. H., *Distributive and modular laws in relation algebras*, Doctoral dissertation, University of California, Berkeley, 1948, 62 pp.

[1951] Chin, L. and Tarski, A., *Distributive and modular laws in the arithmetic of relation algebras*, University of California Publications in Mathematics, new series **1** no. **9** (1951), 341–384.

[1980] Freese, R., *Free modular lattices*, Transactions of the American Mathematical Society **261** (1980), 81–91.

[1994] Givant, S. R., *The structure of relation algebras generated by relativizations*, Contemporary Mathematics vol. 41, American Mathematical Society, Providence, Rhode Island, 1994, xvi + 134 pp.

[1966] Gurevich, Yu., *The word problem for certain classes of semigroups*, Algebra i Logika **5** (1966), 25–35. (Russian)

[1984] Gurevich, Yu., and Lewis, H. R., *The word problem for cancellation semigroups with zero*, Journal of Symbolic Logic **49** (1984), 184–191.

[1971] Henkin, L., Monk, J. D., and Tarski, A., *Cylindric algebras. Part I*, Studies in Logic and the Foundations of Mathematics vol. 64, North-Holland Publishing Company, Amsterdam, 1971, vi + 508 pp.

[1995] Herrmann, C., *On the undecidability of implications between embedded multivalued database dependencies*, Journal of Theoretical Computer Science **122** (1995), 221–235.

[1992] Jipsen, P., *Computer-aided investigations of relation algebras*, Doctoral Dissertation, Vanderbilt University, Nashville, 1992, iii + 82 pp.

[1994] Jipsen, P. and Lukács, E., *Minimal relation algebras*, Algebra Universalis **32** (1994), 189–203.

[1959] Jónsson, B., *Representations of modular lattices and of relation algebras*, Transactions of the American Mathematical Society **92** (1959), 449–464.

[1982] ———, *Varieties of relation algebras*, Algebra Universalis **15** (1982), 273–298.

[1988] ———, *Relation algebras and Schröder categories*, Discrete Mathematics **70** (1988), 27–45.

[1991] ———, *The theory of binary relations*, Algebraic logic (H. Andréka, J. D. Monk, and I. Németi, eds.), Colloquia Mathematica Societatis János Bolyai vol. 54, North-Holland Publishing Company, Amsterdam, 1991, pp. 245–292.

[1951] Jónsson, B. and Tarski, A., *Boolean algebras with operators. Part I*, American Journal of Mathematics **73** (1951), 891–939.

[1952] ———, *Boolean algebras with operators. Part II*, American Journal of Mathematics **74** (1952), 127–162.

[1993] Kurucz, Á., Németi, I., Sain, I., and Simon, A., *Undecidable varieties of semilattice-ordered semigroups, and logics extending the Lambek Calculus*, Bulletin of the Interest Group in Propositional Logic **1** (1993), 91–98.

[1974] Lipshitz, L., *The undecidability of the word problems for projective geometries and modular lattices*, Transactions of the American Mathematical Society **193** (1974), 171–180.

[1961] Lyndon, R. C., *Relation algebras and projective geometries*, Michigan Mathematical Journal **8** (1961), 21–28.

[1978] Maddux, R. D., *Topics in relation algebras*, Doctoral Dissertation, University of California, Berkeley, 1978, iii + 241 pp.

[1981] ———, *Embedding modular lattices into relation algebras*, Algebra Universalis **12** (1981), 242–246.

[1991] ———, *Pair-dense relation algebras*, Transactions of the American Mathematical Society **328** (1991), 83–131.

[1994] ———, *Undecidable semiassociative relation algebras*, Journal of Symbolic Logic **59** (1994), 398–418.

[1995] ———, *Review #03035 of Givant* [1994], Zentralblatt für Mathematik und ihre Grenzgebiete **812** (1995), 31.

[a] Marx, M., Masuch, M., and Polos, L., eds., *Arrow logic and multi-modal logic*, Proceedings of the Conference: Logic at Work, held at the Center for Computer Science in Organization and Management,University of Amsterdam, December 1992, CSLI Publications, to appear.

[1964] Monk, J. D., *On representable relation algebras*, Michigan Mathematical Journal **11** (1964), 207–210.

[1985] Németi, I., *Exactly which varieties of cylindric algebras are decidable?*, Preprint 34/1985, Mathematical Institute of the Hungarian Academy of Sciences, Budapest, 1985, 26 pp.

[1995] Németi, I., Sain, I., and Simon, A., *Undecidability of the equational theories of some classes of residuated Boolean algebras with operators*, Bulletin of the Interest Group in Propositional Logic **3** (1995), 93–107.

[1955] Novikov, P. S., *On the algorithmic unsolvability of the word problem in group theory*, Trudy Matematicheskogo Instituta imeni V. A. Steklova **44** (1955). (Russian)

[1994] Pratt, V. R., *A roadmap of two-dimensional logics*, Logic and Information Flow (J. van Eijck and A. Visser, eds.), MIT Press, Cambridge MA, 1994, pp. 149–162.

[1965] Rabin, M. O., *A simple method for undecidability proofs and some applications*, Logic, Methodology and Philosophy of Science II (Y. Bar-Hillel, ed.), North-Holland Publishing Company, Amsterdam, 1965, pp. 58–68.

[1979] Schönfeld, W., *An undecidability result for relation algebras*, Journal of Symbolic Logic **44** (1979), 111–115.

[1941] Tarski, A., *On the calculus of relations*, Journal of Symbolic Logic **6** (1941), 73–89.

[1946] ———, *A remark on functionally free algebras*, Annals of Mathematics **47** (1946), 163–165.

[1953] ———, *Some metalogical results concerning the calculus of relations*, Journal of Symbolic Logic **18** (1953), 188–189.

[1954] ———, *Contributions to the theory of models. II*, Koninklijke Nederlandse Akademie van Wetenschappen, Proceedings, Series A, Mathematical Sciences, **57** (= Indagationes Mathematicae **16**) (1954), 582–588.

[1955] ———, *Contributions to the theory of models. III*, Koninklijke Nederlandse Akademie van Wetenschappen, Proceedings, Series A, Mathematical Sciences, **58** (= Indagationes Mathematicae **17**) (1955), 56–64.

[1987] Tarski, A. and Givant, S. R., *A formalization of set theory without variables*, American Mathematical Society, Colloquium Publications vol. 41, Providence RI, 1987, xxii + 318 pp.

[1984] Urquhart, A., *The undecidability of entailment and relevant implication*, Journal of Symbolic Logic **49** (1984), 1059–1073.

[1995] ———, *Decision problems for distributive lattice-ordered semigroups*, Algebra Universalis **34** (1995), 399–418.

[1994] van Benthem, J., *Dynamic Arrow Logic*, Logic and Information Flow (J. van Eijck and A. Visser, eds.), MIT Press, Cambridge MA, 1994, pp. 15–29.

[1960] von Neumann, J., *Continuous Geometry*, Princeton Mathematical Series, vol. 25 (I. Halperin, ed.), Princeton University Press, Princeton NJ, 1960, xi + 299 pp.

[1857] von Staudt, K. G. C., *Beiträge zur Geometrie der Lage*, F. Korn, Nürenberg, 1857.

INDEX OF SYMBOLS

Set-theoretic notions

$x \in X$	x is a member of X, 1
$x \notin X$	x is not a member of X, 1
$\varnothing$	empty set, 3
$\{x, y, \dots\}$	set of elements $x, y, \dots$, 7
$\{x : \varphi[x]\}$	set of all elements satisfying φ, 1
(x, y)	ordered pair of x and y, 1
$\langle a_i : i < n \rangle$	sequence of the elements $a_0, \dots, a_{n-1}$, 12
$a^\frown b$	concatenation of the sequences a and b, 1
$X \subseteq Y$	X is a subset of Y, 8
$Sb\,(X)$	set of all subsets of X, 1
$X \cup Y$	union of X and Y, 27
$X \cap Y$	intersection of X and Y, 8
$\bigcup X$	union of all members of X, 4
$\bigcap X$	intersection of all members of X, 63
$\bigcup_{i \in I} X_i$	union of family of sets X_i, 63
$\bigcap_{i \in I} X_i$	intersection of family of sets X_i, 63
$X \sim Y$	difference of X and Y, 1
$\sim Y$	complement of Y in a fixed universe, 1
$|X|$	cardinality of the set X, 1
$X \times Y$	set of ordered pairs (x, y) with $x \in X$ and $y \in Y$, 1
$^X Y$	set of all functions from X to Y, 1
$\prod_{n \in \omega} X_n / F$	ultraproduct of X_n modulo the ultrafilter F, 42
$f[X]$	image of X under the function f, 1
$f^{-1}[Y]$	inverse image of Y under the function f, 1
P, Q, R, S, T	usually denote binary relations, 1
I_U	identity relation on the set U, 1
D_U	diversity relation on the set U, 1
I	I_U when U is understood, 1
D	D_U when U is understood, 1
$R \mid S$	relational composition of R and S, 1
$F \circ G$	functional composition of F and G, 1
R^{-1}	relational inverse of R, 1

Number theoretic notions

$0, 1, 2, \ldots$	natural numbers, 1
∞	infinity symbol, 67
$-\infty$	negative infinity symbol, 67
ω	set of natural numbers, 1
$\hat{\omega}$	$\omega \sim \{0, 1, 2\}$, 62
X'	the set $\{n + 3 : n \in X\}$, 65
Z	set of integers, 30
Z_e	set of extended integers $Z \cup \{-\infty, \infty\}$, 80
$Cf(Z)$	collection of finite and cofinite subsets of Z, 67
$[k, l]$	closed interval $\{n \in Z : k \leq n \leq l\}$, 67
(k, l)	open interval $\{n \in Z : k < n < l\}$, 67
$[k, \infty)$	closed interval $\{n \in Z : k \leq n\}$, 67
$(-\infty, k)$	open interval $\{n \in Z : n < k\}$, 67
$d \mid n$	n is divisible by d, 100
$\lfloor x \rfloor$	greatest integer $\leq x$, 1
$\lceil x \rceil$	smallest integer $\geq x$, 1
$m \equiv n \mod d$	congruence of integers mod d, 99
$n!$	n factorial, 99
$x + y$	sum of numbers x and y, 30 sum of extended integers, 67
$+(x, y, z)$	ternary relation $z = x + y$, 80
$\sum_{i<n} x_i$	sum of the integers $x_0, \ldots, x_{n-1}$, 99
$x - y$	difference of numbers x and y, 1
$-x$	negative of number x, 30
$x \cdot y$	product of the numbers x and y, 32
x/y	rational quaotient of integers x and y, 101
$x \leq y$	inequality between numbers x and y, 21
$x < y$	strict inequality between numbers x and y, 9
R_m	relation of equality between m-ary sums, 99
S_m	relation of inequality between m-ary sums, 99
T_d	relation of divisibility of an m-ary sum by d, 100
$r_m(x)$	remainder of x upon division by m, 101
P, p, N, n, z	indicator functions, 75
$P(x), p(x), N(x),$ $n(x), z(x)$	indicators of x, 75
$P_i, p_i, N_i,$ n_i, z_i	indicators of x_i, 76
In_x	set of indicators of x, 76

Algebraic notions

1'	identity element, 2
1	Boolean unit, 3
0	Boolean zero, 3
0'	diversity element, 3

$\lvert x, y \rvert_j$	length of minimal path from x to y in $\mathfrak{U}_j$, 72
$[\mathsf{M}, \mathsf{N}]$	lattice intervalbetween M and N, 7
$\simeq$	an equivalence between integers, 95

Classes of algebras

$\mathsf{K}, \mathsf{L}, \ldots$	classes of algebras, 1
$\mathbf{H}(\mathsf{K})$	homomorphic images of algebras in K, 1
$\mathbf{S}(\mathsf{K})$	isomorphic images of subalgebras of algebras in K, 1
$\mathbf{P}(\mathsf{K})$	isomorphic images of direct products of algebras in K, 1
$\mathbf{Si}(\mathsf{K})$	simple algebras in K, 6
RA	relation algebras, 3
RRA	representable relation algebras, 4
GRA	algebras isomorphic to group relation algebras, 5
I_n	simple relation algebras of integrality degree n, 44
RA_1	positive complement-converse reducts, 45
L	algebras embeddable into Lyndon algebras, 56
L_X	$\mathsf{L} \sim \mathsf{F}_X$, 62
F_X	$\{\mathfrak{A} : \mathfrak{A} \cong \mathfrak{Ln}(n)$ for some n in $\sim X\}$, 62

Logical notions

$\sigma, \tau, \varphi, \psi, \ldots$	terms or formulas, 2
$\varphi \wedge \psi$	conjunction of φ and ψ, 6
$\bigwedge_{i<n} \varphi_i$	conjunction of $\varphi_0, \ldots, \varphi_{n-1}$, 9
$\varphi \to \psi$	implication from φ to ψ, 6
$\varphi \leftrightarrow \psi$	equivalence of φ and ψ, 6
$\forall x \varphi$	universal quantification of φ, 6
$\tau(\sigma_0, \ldots, \sigma_{n-1})$	simultaneous substitution of $\sigma_0, \ldots, \sigma_{n-1}$ into τ, 2
$\sigma(x)$	formula defining singleton subdiversity elements, 56
ε_φ	equation associated with φ, 6
$\varphi_\mathfrak{A}$	sentence expressing embeddability of $\mathfrak{A}$, 7
$\tau^\mathfrak{A}, \tau_n^\mathfrak{A}$	operation induced in $\mathfrak{A}$ by τ, 2
$\tau^\mathfrak{A}[a], \tau[a]$	value of $\tau^\mathfrak{A}$ at a, 2
$\tau^\mathfrak{A}[v_0, \ldots, v_{n-1}, b]$	operation induced in $\mathfrak{A}$ by τ with parameters b, 2
$\mathfrak{A} \models \varphi[a]$	a satisfies the formula φ in $\mathfrak{A}$, 2
$\varphi^\mathfrak{A}$	relation induced in $\mathfrak{A}$ by φ, 2
$\varphi^\mathfrak{A}[v_0, \ldots, v_{n-1}, b]$	relation induced in $\mathfrak{A}$ by φ with parameters b, 2
$\mathcal{L}_r$	first-order language of relation algebras, 3
$\mathcal{E}q(\mathsf{K})$	equational theory of K, 2
Λ	set of axioms for L, 56
g_n, h_{mn}	recursive functions, 107

INDEX OF NAMES AND SUBJECTS

To keep the number of page references for any given item down to a usable size, the index contains no explicit references to terms when they are used in proofs. However, it does contain references to authors whose works are cited in proofs. A boldface number indicates a principal reference, or a reference to a definition.

H

I

J

S

Editorial Information

To be published in the *Memoirs*, a paper must be correct, new, nontrivial, and significant. Further, it must be well written and of interest to a substantial number of mathematicians. Piecemeal results, such as an inconclusive step toward an unproved major theorem or a minor variation on a known result, are in general not acceptable for publication. *Transactions* Editors shall solicit and encourage publication of worthy papers. Papers appearing in *Memoirs* are generally longer than those appearing in *Transactions* with which it shares an editorial committee.

As of September 30, 1996, the backlog for this journal was approximately 7 volumes. This estimate is the result of dividing the number of manuscripts for this journal in the Providence office that have not yet gone to the printer on the above date by the average number of monographs per volume over the previous twelve months, reduced by the number of issues published in four months (the time necessary for preparing an issue for the printer). (There are 6 volumes per year, each containing at least 4 numbers.)

A Copyright Transfer Agreement is required before a paper will be published in this journal. By submitting a paper to this journal, authors certify that the manuscript has not been submitted to nor is it under consideration for publication by another journal, conference proceedings, or similar publication.

Information for Authors and Editors

Memoirs are printed by photo-offset from camera copy fully prepared by the author. This means that the finished book will look exactly like the copy submitted.

The paper must contain a *descriptive title* and an *abstract* that summarizes the article in language suitable for workers in the general field (algebra, analysis, etc.). The *descriptive title* should be short, but informative; useless or vague phrases such as "some remarks about" or "concerning" should be avoided. The *abstract* should be at least one complete sentence, and at most 300 words. Included with the footnotes to the paper, there should be the 1991 *Mathematics Subject Classification* representing the primary and secondary subjects of the article. This may be followed by a list of *key words and phrases* describing the subject matter of the article and taken from it. A list of the numbers may be found in the annual index of *Mathematical Reviews*, published with the December issue starting in 1990, as well as from the electronic service e-MATH [**telnet e-MATH.ams.org** (or **telnet 130.44.1.100**). Login and password are **e-math**]. For journal abbreviations used in bibliographies, see the list of serials in the latest *Mathematical Reviews* annual index. When the manuscript is submitted, authors should supply the editor with electronic addresses if available. These will be printed after the postal address at the end of each article.

Electronically prepared papers. The AMS encourages submission of electronically prepared papers in AMS-TeX or AMS-LaTeX. The Society has prepared author packages for each AMS publication. Author packages include instructions for preparing electronic papers, the *AMS Author Handbook*, samples, and a style file that generates the particular design specifications of that publication series for both AMS-TeX and AMS-LaTeX.

Authors with FTP access may retrieve an author package from the Society's Internet node `e-MATH.ams.org` (130.44.1.100). For those without FTP access, the author package can be obtained free of charge by sending e-mail to `pub@math.ams.org` (Internet) or from the Publication Division, American Mathematical Society, P.O. Box 6248, Providence, RI 02940-6248. When requesting an author package, please specify $\mathcal{A}_{\mathcal{M}}\mathcal{S}$-TEX or $\mathcal{A}_{\mathcal{M}}\mathcal{S}$-LATEX, Macintosh or IBM (3.5) format, and the publication in which your paper will appear. Please be sure to include your complete mailing address.

Submission of electronic files. At the time of submission, the source file(s) should be sent to the Providence office (this includes any TEX source file, any graphics files, and the DVI or PostScript file).

Before sending the source file, be sure you have proofread your paper carefully. The files you send must be the EXACT files used to generate the proof copy that was accepted for publication. For all publications, authors are required to send a printed copy of their paper, which exactly matches the copy approved for publication, along with any graphics that will appear in the paper.

TEX files may be submitted by email, FTP, or on diskette. The DVI file(s) and PostScript files should be submitted only by FTP or on diskette unless they are encoded properly to submit through e-mail. (DVI files are binary and PostScript files tend to be very large.)

Files sent by electronic mail should be addressed to the Internet address `pub-submit@math.ams.org`. The subject line of the message should include the publication code to identify it as a Memoir. TEX source files, DVI files, and PostScript files can be transferred over the Internet by FTP to the Internet node `e-math.ams.org` (130.44.1.100).

Electronic graphics. Figures may be submitted to the AMS in an electronic format. The AMS recommends that graphics created electronically be saved in Encapsulated PostScript (EPS) format. This includes graphics originated via a graphics application as well as scanned photographs or other computer-generated images.

If the graphics package used does not support EPS output, the graphics file should be saved in one of the standard graphics formats—such as TIFF, PICT, GIF, etc.—rather than in an application-dependent format. Graphics files submitted in an application-dependent format are not likely to be used. No matter what method was used to produce the graphic, it is necessary to provide a paper copy to the AMS.

Authors using graphics packages for the creation of electronic art should also avoid the use of any lines thinner than 0.5 points in width. Many graphics packages allow the user to specify a "hairline" for a very thin line. Hairlines often look acceptable when proofed on a typical laser printer. However, when produced on a high-resolution laser imagesetter, hairlines become nearly invisible and will be lost entirely in the final printing process.

Screens should be set to values between 15% and 85%. Screens which fall outside of this range are too light or too dark to print correctly.

Any inquiries concerning a paper that has been accepted for publication should be sent directly to the Editorial Department, American Mathematical Society, P. O. Box 6248, Providence, RI 02940-6248.

Selected Titles in This Series

(Continued from the front of this publication)

573 **Wayne Aitken,** An arithmetic Riemann-Roch theorem for singular arithmetic surfaces, 1996
572 **Ole H. Hald and Joyce R. McLaughlin,** Inverse nodal problems: Finding the potential from nodal lines, 1996
571 **Henry L. Kurland,** Intersection pairings on Conley indices, 1996
570 **Bernold Fiedler and Jürgen Scheurle,** Discretization of homoclinic orbits, rapid forcing and "invisible" chaos, 1996
569 **Eldar Straume,** Compact connected Lie transformation groups on spheres with low cohomogeneity, I, 1996
568 **Raúl E. Curto and Lawrence A. Fialkow,** Solution of the truncated complex moment problem for flat data, 1996
567 **Ran Levi,** On finite groups and homotopy theory, 1995
566 **Neil Robertson, Paul Seymour, and Robin Thomas,** Excluding infinite clique minors, 1995
565 **Huaxin Lin and N. Christopher Phillips,** Classification of direct limits of even Cuntz-circle algebras, 1995
564 **Wensheng Liu and Héctor J. Sussmann,** Shortest paths for sub-Riemannian metrics on rank-two distributions, 1995
563 **Fritz Gesztesy and Roman Svirsky,** (m)KdV solitons on the background of quasi-periodic finite-gap solutions, 1995
562 **John Lindsay Orr,** Triangular algebras and ideals of nest algebras, 1995
561 **Jane Gilman,** Two-generator discrete subgroups of $PSL(2, R)$, 1995
560 **F. Tomi and A. J. Tromba,** The index theorem for minimal surfaces of higher genus, 1995
559 **Paul S. Muhly and Baruch Solel,** Hilbert modules over operator algebras, 1995
558 **R. Gordon, A. J. Power, and Ross Street,** Coherence for tricategories, 1995
557 **Kenji Matsuki,** Weyl groups and birational transformations among minimal models, 1995
556 **G. Nebe and W. Plesken,** Finite rational matrix groups, 1995
555 **Tomás Feder,** Stable networks and product graphs, 1995
554 **Mauro C. Beltrametti, Michael Schneider, and Andrew J. Sommese,** Some special properties of the adjunction theory for 3-folds in $\mathbb{P}^5$, 1995
553 **Carlos Andradas and Jesús M. Ruiz,** Algebraic and analytic geometry of fans, 1995
552 **C. Krattenthaler,** The major counting of nonintersecting lattice paths and generating functions for tableaux, 1995
551 **Christian Ballot,** Density of prime divisors of linear recurrences, 1995
550 **Huaxin Lin,** C^*-algebra extensions of $C(X)$, 1995
549 **Edwin Perkins,** On the martingale problem for interactive measure-valued branching diffusions, 1995
548 **I-Chiau Huang,** Pseudofunctors on modules with zero dimensional support, 1995
547 **Hongbing Su,** On the classification of C*-algebras of real rank zero: Inductive limits of matrix algebras over non-Hausdorff graphs, 1995
546 **Masakazu Nasu,** Textile systems for endomorphisms and automorphisms of the shift, 1995
545 **John L. Lewis and Margaret A. M. Murray,** The method of layer potentials for the heat equation on time-varying domains, 1995
544 **Hans-Otto Walther,** The 2-dimensional attractor of $x'(t) = -\mu x(t) + f(x(t-1))$, 1995
543 **J. P. C. Greenlees and J. P. May,** Generalized Tate cohomology, 1995

(See the AMS catalog for earlier titles)